DESIGN AND BUILD A GREAT
18th CENTURY ROOM

JEFF KNUDSEN

4880 Lower Valley Road, Atglen, PA 19310 USA

Dedication

In memory of my father, Heinrich N. Knudsen, a man of great gentleness and integrity.

Library of Congress Cataloging-in-Publication Data

Knudsen, Jeff.
  Design and build a great 18th century room/Jeff Knudsen.
    p. cm.
  ISBN 0-7643-1423-8 (hardcover)
  1. Dwellings--Remoldeling. 2. Carpentry. 3. Architecture,
Georgian--United States. I. Title
  TH4816 .K598 2001
  694'.6--dc21

                                   2001003208

Designed by John P. Cheek
Cover design by Bruce M. Waters
Type set in Americana XBd BT/Zapf Humanist BT

ISBN: 0-7643-1423-8
Printed in China

Published by Schiffer Publishing Ltd.
4880 Lower Valley Road
Atglen, PA 19310
Phone: (610) 593-1777; Fax: (610) 593-2002
E-mail: Schifferbk@aol.com
Please visit our web site catalog at
www.schifferbooks.com
We are always looking for people to write books on new and related subjects. If you have an idea for a book please contact us at the above address.

This book may be purchased from the publisher.
Include $3.95 for shipping.
Please try your bookstore first.
You may write for a free catalog.

In Europe, Schiffer books are distributed by
Bushwood Books
6 Marksbury Ave.
Kew Gardens
Surrey TW9 4JF England
Phone: 44 (0)20-8392-8585
Fax: 44 (0)20-8392-9876
E-mail: Bushwd@aol.com
Free postage in the UK. Europe: air mail at cost

Contents

Acknowledgments

Like most cabinetry, this book was done with a lot of help from my friends. A sincere thank you to these folks:

My wonderful wife Domenica for her support and patience through this project.

Shane Knudsen, who contributed all the drawings and helped with many other parts of this book.

Stephanie Hanna, whose editing skills made sense of a sometimes less than articulate manuscript.

The following individuals, who lent their time and skills to this book: Seth Botkin, Chris Ferrier, Ted French, Steve Latta, Joe Neuman, John Taylor, "Woody" Woodring, and Ben Parry.

Lastly, a special thanks to the man who has taught me most of what I know about cabinetry and much of what I know about integrity – John Bland, the finest cabinetmaker I have ever known.

Foreword

All too often we are reminded that we live in an age of compromise. We settle for this; we put up with that – particularly in matters involving craftsmanship. Past examples of truly exceptional work are plentiful: Fine old houses display our country's best architectural design, carpentry, furniture, glasswork, and so on. Ironically, these standards were established in a much simpler time, and, despite our technological advancements, we regularly fail to meet them. We look at old rooms with rich paneling and incredible moldings and, in our attempts to reproduce such a "feeling" in our own work, fall short of the subtle grandeur those rooms achieved. Typically, our efforts fail for two reasons – lack of woodworking skills and lack of design knowledge.

In our fast-paced world, instant gratification with minimal effort is the norm. Similarly, in our reproductions of things past, we look for simple equations and design formulas. However, while such formulas do often surface, they only appear after hours of study. Rooms, with their subtle nuances of layout and proportion, require special study. An understanding of why some rooms work well and others don't requires more experience than a quick glance at a few pictures in books and a stroll through a few actual rooms affords. To know the genre requires study. It demands a time investment, touring multiple homes, and sitting and absorbing the feeling each room offers before moving to measuring tape and note pad. Without such effort, our designs lack the sophistication necessary for rooms to actually become what we intend.

Often, our technical aptitude compromises the design. Our shift in production techniques from hands to machines has sacrificed the ability to produce many wonderful details. The uniform results of machined products work well in many circumstances, but they lack the richness of hand-planed panels or carved mantles. The strength of the joinery, the complexity of the moldings, and other detailing all blend to make rooms succeed. As cabinetmakers, our machine and hand skills must be strong enough to meet the demands of the design.

This book blends a thorough understanding of both design and cabinetmaking skills. It is a byproduct of countless hours spent examining pictures and blueprints of Georgian rooms and even more hours visiting rooms up and down the East Coast, measuring, taking notes, and simply having the patience to sit long enough to get the true feeling of particular rooms.

This is not a book for novice woodworkers or cabinetmakers. Jeff assumes that readers both know their way around a shop and have the technical abilities to take an obscure, undefined idea and turn it into reality. This book deals more with idea and design than with process. It is a wonderful starting point for anyone wanting to design and build a room with eighteenth century flavor. It is only a beginning, however. As Jeff points out repeatedly, readers must take this knowledge with them as they formulate their own ideas, studying photo after photo and visiting room after room.

– Steve Latta

A Word About Safety

Creating fine furniture and cabinetry in a medium as beautiful as wood is quite satisfying, but can be dangerous as well. Your own common sense is the best safety device in the world. If an operation does not feel rock-solid, re-think it. Slow down. The consequences of a bad setup can be disastrous.

Use the guards provided on your tools. We had to remove some of the guards for photographic clarity – you don't. A properly installed guard can make all the difference in safety.

Enjoy your craft to its fullest extent, but always keep safety first.

Chapter 1
Design

History of Design

We do not intend this book to be an academic history of design, but a practical how-to book enabling you to design a beautiful room. However, a brief background of the evolution of Georgian style benefits any designer.

In the late 1600s, western European architects began making pilgrimages to what are today Italy and Greece, collecting examples of shapes and details from ancient architecture, and hoping to transform elements from those stone structures into contemporary eighteenth century stone and wooden buildings. Similarly, to design within this genre, you should first experience it, doing exactly what 17th century architects did – visit old houses. Visiting old houses, particularly those staffed by knowledgeable tour guides, affords many opportunities to ask questions and learn experientially. Many of the houses also have detail drawings available for purchase; these are quite helpful as you view the moldings and cabinets before you. This method of shape and detail gathering applies equally well to other periods of design.

Books provide another important source of information. Museum stores and book stores sell many helpful books on interior architecture. Also, used book stores often have out-of-print books and others at a fraction of the new book cost. But take care that the photographic quality is sufficient, with enough close-ups and details to be useful to a cabinetmaker. (See appendix.)

Designing a Room

Rarely when designing rooms do we have the luxury of designing the entire space. More often, walls, windows, and doors exist previously, and we must adapt our design to work with them.

Also, customers have requirements for rooms' functions (as book storage, TV viewing areas, business meeting spaces, etc.). Before making in-depth drawings, make certain that the room can adequately accommodate these needs, or that the customer is willing to compromise some functions if necessary.

After gathering this functional information, decide on the level of style appropriate for the room. Eighteenth century style can vary from primitive farmhouses to sophisticated high-style townhouses with lots of Rococo carvings. The level of style may also change somewhat

according to the room's importance. However, the style should be generally consistent with the exterior of the house. For instance, a house with a detailed and formal exterior should never have a primitive parlor, and the interior of a simple stone farmhouse should not be done in high style.

Each room presents its own set of challenges for the designer. Further, each wall must be designed separately, then considered in relation to the others. This creates a cohesive look that incorporates the various functions required by the client.

In this section, we will proceed through the design process of several rooms step-by-step, from the customer's initial requirements to the final room design.

Room #1

The client needed a room that combined book storage, display shelving, a television, and general storage. Aesthetically, the client requested a hardwood paneled private study that would in some way reflect his interests, particularly his interest in upland game hunting.

After looking at a variety of wood samples and color options, the customer chose walnut because of its depth and dramatic grain pattern. Once the customer's wishes are known, next decide on the level of style. Choice of style is somewhat subjective in nature, but the room design should reflect the overall level of detail in the house.

Because this room is a private study, the level of detail should be more elaborate than that of a bedroom or another less important room. Once you've established the general level of style, choose the panel detail that will reflect that style. In Room #1, we selected a beaded panel with an ovolo sticking. This combination always looks great and firmly sets the style of the room.

Once you've determined the panel detail, turn your attention to the most important fixture in the room. If there is a fireplace in the room, design this next. This room had no fireplace, so we turned our attention to the bookcases.

You can divide a wall into properly proportioned bookcases in a number of ways, some better than others. Visiting old houses and consulting a stack of reference books is key to deciding which type of cabinetry will best suit the space. Choose the style that is most pleasing, fits in the space, and conforms to the level of style and period design of the house.

The process of fitting a design to a wall can be frustrating. Often, cabinetry fits in your mind's eye but doesn't work on paper. If a particular design can't work on your wall, give it up and move on to another design. It's better to have a well-proportioned second choice than to use your first choice and have it be out of proportion to the room.

When the paneling and cabinet style are decided, move on to the molding details, the most prominent of which is the cornice. The most important aspect of a cornice is its size. A workable "Rule of Thumb" is to make the height of the cornice one-fifteenth the height of the room. Guidelines such as this can be very helpful, but don't be a slave to them. For example, in Room #1 the cornice is only 6" high, though the room height is 10 feet. We made the cornice about one-twentieth of the room height in this case because it has a soffit which extends the cornice further into the room, making it appear larger. The other moldings in the room also work well together, their size and patterns being based on eighteenth century moldings rather than twentieth century lumberyard patterns.

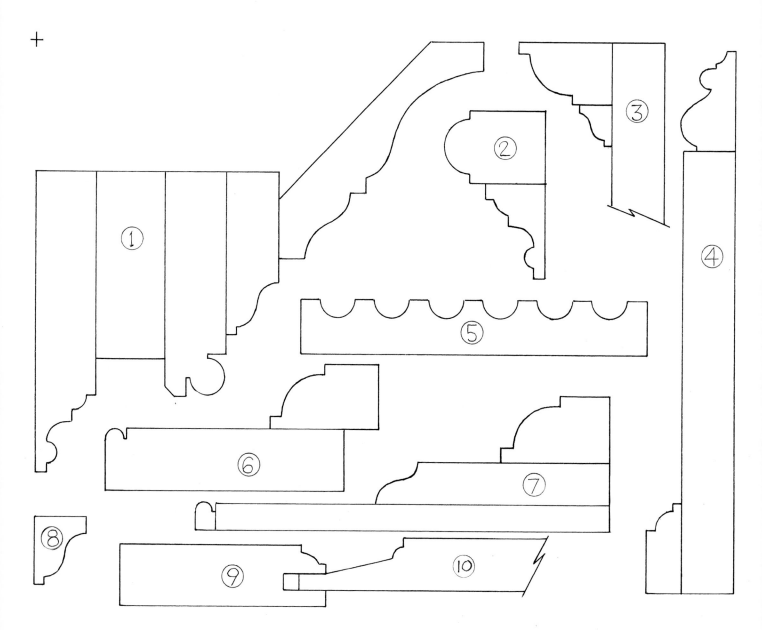

Molding details of Room #1: 1) cornice; 2) mid-molding; 3) arched pilaster cap; 4) baseboard; 5) pilaster; 6) casing; 7) arched casing; 8) small pilaster cap; 9) style and rail sticking; 10) panel detail.

Hopefully the preceding pages provide a good start on the journey to becoming a designer.

Think of learning period design as being similar to learning a foreign language. Each language has its own sentence structures and peculiarities, just as each archi-tectural period has peculiar set of spatial relationships and design elements. Like learning a language, the best way to understand an architectural period is to immerse yourself in it.

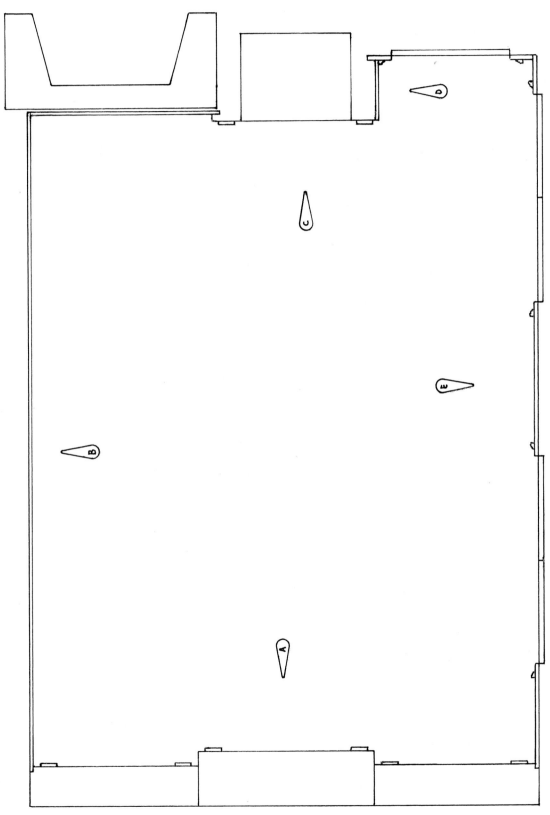

Footprint of Room #1.

Elevation A

We chose to satisfy most of the room's physical requirements in Elevation A. The wall's length allowed space for three nicely proportioned cabinets, each including pilaster strips. The center cabinet, deeper than the two flanking cabinets, offers display space, and the raised panel doors below each cabinet meet the storage requirements.

Note that in these original drawings, the pilasters also had fan carvings. However, during one design meeting, the customer asked if we could somehow personalize the room. Hence, we added upland bird carvings to reflect his interest. Details like these, though not strictly eighteenth century, are fun to design and add greatly to customer satisfaction. ·

Elevation A, bookcases with storage underneath.

Elevation A.

Elevation B

Elevation B is an interior wall whose only obstructions are several electrical outlets and a cold air return. Usually when designing a wall like this, we choose a panel size proportional to the room and one that we can most closely duplicate in the rest of the room's walls. However, this particular customer has a great appreciation for figured hardwoods, so we used our best crotch walnut flitch, the width of which determined the width of the room's paneling.

Elevation B, matched panels, sequential as they were sawed from the log.

In design, the electrical outlets almost always look best in the baseboard. However, heat and cold air returns are sometimes difficult to incorporate into an eighteenth century room. In this case, we chose to slot a panel to allow for the air return. Before installing paneling, we sprayed the duct work with black paint so that the duct work can't be seen through the panel.

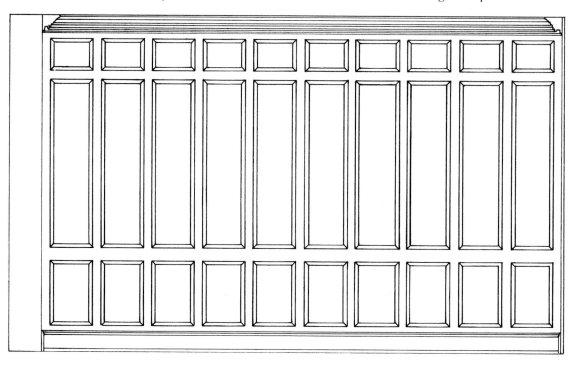

Elevation B.

Elevation C

Elevation C proved the most difficult of the four walls to design because it had to accommodate both a TV and a door opening. Further, a fireplace in the adjacent room took up the left 1/3 of the wall. We thus designed the TV cabinet to occupy a space that previously held a cabinet in the adjacent room.

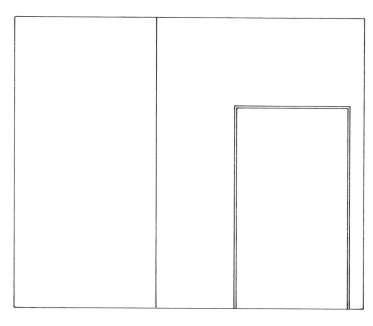

Elevation D

We show Elevation D simply to illustrate that all cabinet sides should be paneled similarly to walls and cabinet fronts. Whenever possible, panels on the sides of cabinets should line up horizontally with other panels in the room.

Elevation C, existing wall.

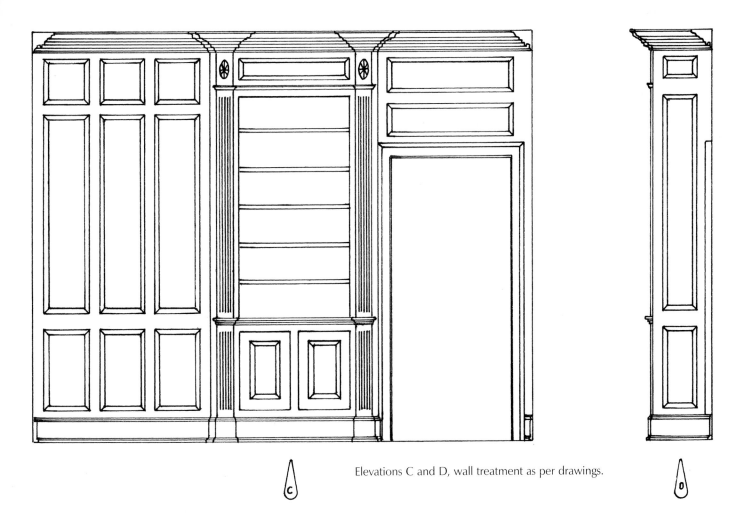

Elevations C and D, wall treatment as per drawings.

Elevations C and D completed. Note changes made after completed drawings: doors were added, and carvings changed to personalize room.

Elevation E

Walls like this one serve as reality checks when design is going well: The designer cannot bring symmetry to the paneling on this wall. Two French doors, not centered in the room, dominate Elevation E. Moving the doors was not an option, as it would negatively affect the exterior of the building. In such a case, designers can only make the panels as close in size as possible.

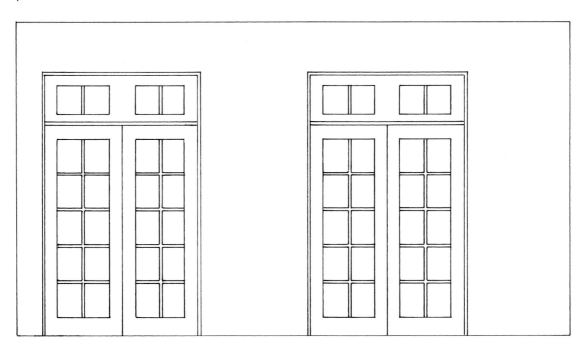

Elevation E, existing French doors.

Elevation E, paneling treatment.

Elevation E.

Room #2

The customer wished to use this room to store clothing and miscellaneous items, choosing the shaker style because of its simplicity, beauty, and efficiency. Shaker design is not an eighteenth century style of cabinetry, but we include it here to demonstrate that the basic design principals of one architectural period transfer easily to another. (Dozens of other books document shaker interior architecture in detail.)

As you familiarize yourself with shaker style's basic shapes and connections between shapes, a look should begin to emerge in your mind's eye. First, you'll notice the shaker penchant for simplicity. (This design simplicity grew out of the shakers' efforts to achieve a simple lifestyle, free from unnecessary embellishment.)

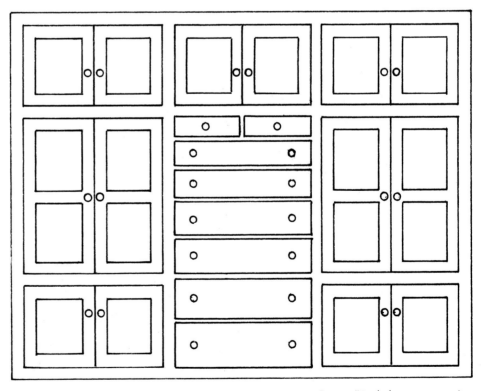

Room #2, shaker storage unit.

Next, you'll notice the choice of shape – generally rectilinear, and more square than eighteenth century design. The shakers also had a strong sense of symmetry, but would deviate from that symmetry for the sake of practicality. This simplicity of design accentuates the importance of the wood chosen for shaker-style work. Most commonly, shakers used cherry, white pine, and maple.

Shaker storage unit
in figured cherry.

Room #3

The customer's requirements for this room were fairly simple: create a beautiful wall with fireplace and display capabilities. This house, under construction during the design process, faithfully reproduces eighteenth century design, and the client, an avid antiques collector, knew a lot about eighteenth century details. Her knowledge, enthusiasm, and stack of magazine clippings made it easy to draw this wall. Clients who know exactly what they want speed up the design process and make it more pleasurable for all involved.

This room's challenge was to correctly match the level of style already set by the house's design. We accomplished this by paying close attention to the house's design details and carefully considering all local and regional influences.

Shell cabinets and fireplace wall, Room #3.

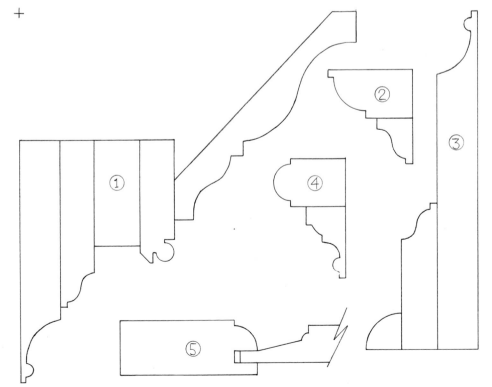

Molding detail, cornice base and sills, Room #3: 1) cornice; 2) pilaster cap; 3) cabinet base; 4) cabinet mid molding; 5) sticking and panel detail.

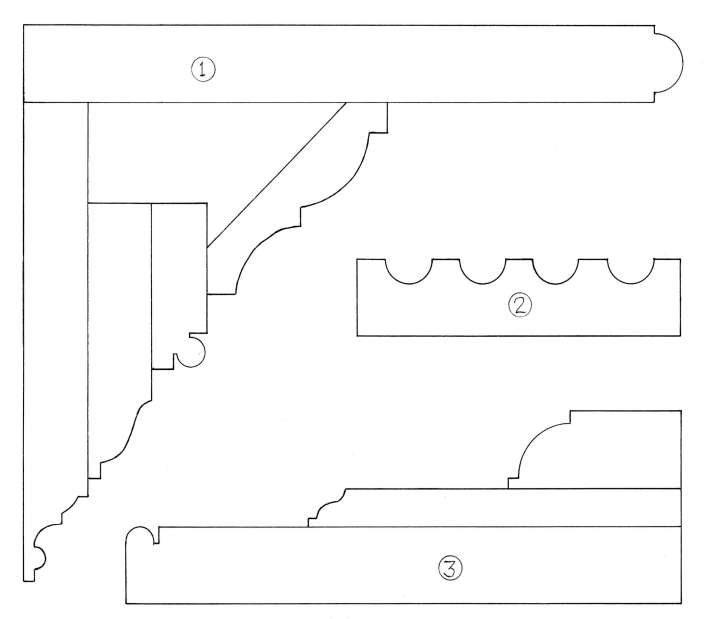

Molding detail, fireplace, Room #3: 1) mantle; 2) pilaster; 3) surround.

Room #3 completed.

Room #4

This customer required display area, book storage, and general storage in the room. Here, we were challenged to incorporate a large window into the overall wall design. We did so by using panels and moldings, which connected the window style to the cabinetry style. We also fitted the room with wainscot all around, connecting the wall aesthetically to the rest of the room.

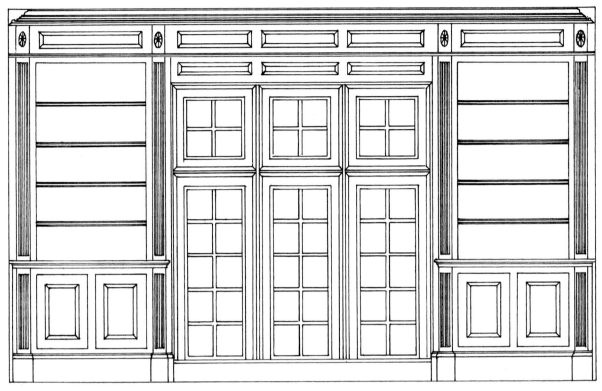

Room #4, completed drawing.

Room #4 completed.

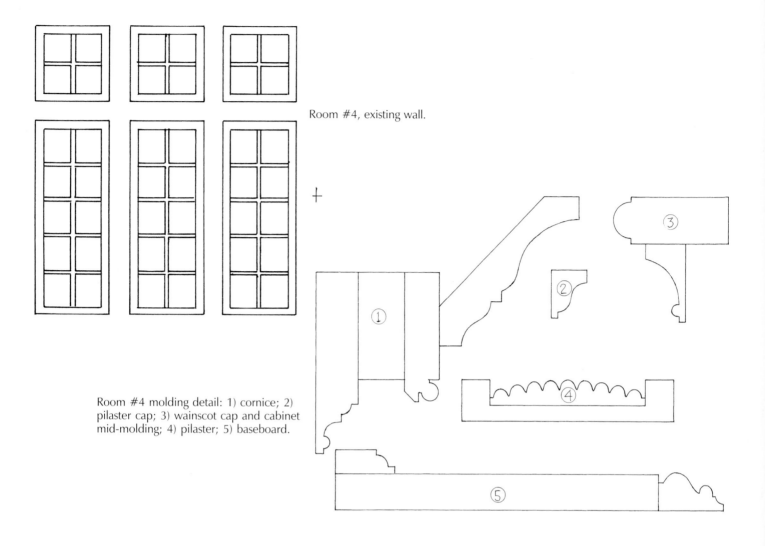

Room #4, existing wall.

Room #4 molding detail: 1) cornice; 2) pilaster cap; 3) wainscot cap and cabinet mid-molding; 4) pilaster; 5) baseboard.

Drawing Materials

When most cabinetmakers are hired to create rooms, they must work within existing parameters, whether because they are remodeling previously constructed rooms or working from architects' drawings of rooms not yet constructed. In the first case, go to the job site and take accurate measurements of the room, including all walls, windows, heating ducts, etc. Often, thumbnail sketches, or rough drawings, are made with the numbers drawn right in. Other thumbnails stem from your, or your customers', possible ideas.

Though it is possible to build cabinetry from rough drawings or even thumbnail sketches, as a rule, making good, accurate drawings before applying a saw to wood will greatly benefit you over time. On a good drawing, all cabinet sizes are figured, all moldings defined, and exact distance relationships noted. Such drawings may seem to be a lot of work, but approached methodically, they become manageable. Above all, a draftsman must have a good supply of patience: drawings take time. For example, drawings of the walnut study (Room #1 detailed in this chapter) required an entire day and a half. This seems like a substantial amount of time, but an unexpected error could cost much more. Additionally, good drawings can be shown to customers: Expensive jobs are much easier to sell when customers can see exactly what they are getting and understand the work behind the costs.

Cabinetmakers generally print handmade drawings as blueprints for use on jobs and keep the pencil originals safely in storage. To make them easy to photocopy, do drawings for blueprints on translucent vellum with a dark pencil. HB lead in a 0.5 mm pencil offers both clear lines and dark lead. Harder leads can be too light, and softer leads can be unclear.

Equipment for drawing is fairly elementary. You can use a simple, flat drawing board with a moveable horizontal member, or a T-square; or, if you plan a lot of drawing, you can purchase new or used dedicated drawing boards with rotating, scaled, square heads.

Necessary drawing tools, left to right: 1) triangles; 2) protractor; 3) adjustable triangle; 4) pen-style eraser; 5) lead; 6) mechanical pencil; 7) engineer's scale; 8) divider; 9) compasses; 10) brush; 11) magnifying glass; 12) adjustable curve; 13) eraser; 14) eraser shield; 15) circle template; 16) arrow template.

Drawing board.

In addition to a board, necessary items include scale rules, transparent squares, high quality erasers, eraser shields, and a dusting brush. Helpful extras include compasses or a circle template, an adjustable angle square, a French curve set, a magnifying glass, and a protractor.

All of these, with a simple drawing board, should cost between $200 and $300.

Scale drawings are essential in the design and construction of any room. Drawings allow you to move ideas from concepts to reality. In them, you figure actual sizes of carcasses and spell out the molding styles. Once complete, you can use the drawings as references during any construction phase.

Chapter Two
Material Choice

The style of the room, amount of available light, and your customer's personal wood preferences all influence your choice of wood for particular rooms, making wood choice always subjective. However, existing eighteenth century rooms and furniture can help to guide the decision making process. Most eighteenth century rooms were painted or left unfinished. However, some notable exceptions of finished hardwood and softwood rooms exist scattered throughout the original thirteen colonies.

Once again, you can best make these choices by visiting old houses and using reference books. For example, if a customer wants to panel a room with mahogany paneling and lots of rococo detail, but the room is in a primitive Pennsylvania farmhouse (circa 1750), the style of the house dictates a much simpler approach, such as hand-planed painted pine. Doing this room in mahogany would be a disservice to the old house and to your credibility as a designer.

Therefore, you should next try to "educate" your customer in wood design. Results at this endeavor can vary from appreciative, satisfied customers to complete strike-outs. In the case of the latter, if the work really doesn't fit the house, it's probably best to decline the job.

Further, when selecting wood, consider the amount of available light in the room. Painted wood tends to reflect light around the room, whereas finished wood seems to absorb light.

Painted Wood

Since most eighteenth century rooms were painted, cabinetmakers chose wood based mainly on availability and cost. The same holds true today. The best wood for a painted project has a closed-grain cell structure and is available in lengths and widths necessary for the job. For example, in the mid-Atlantic region, cabinetmakers most often use poplar and soft maple.

Manmade Materials

Due to the natural seasonal changes in wood, we sometimes choose to use manmade materials in certain situations, specifically for painted carcass parts and raised panels. In the eighteenth century, shrinkage and expansion were considered natural traits of the wood. Today,

however, customers are less likely to accept these attributes. Therefore, we often use cabinet grade plywood for interior carcass parts and medium density fiberboard (MDF) for raised panels. MDF comes in many different grades, the best of which is made of the most finely ground particulate matter. High quality MDF costs about $25 a sheet, paint grade plywood about $35 a sheet, and up to $90 a sheet for finish grade walnut, cherry, and mahogany plywood.

Mahogany

Mahogany and mahogany crotches were reserved for the most sophisticated furniture of the eighteenth century. Imported from Cuba and its surrounding islands at great expense, this dense and strikingly beautiful Caribbean wood looks substantially nicer than the mahogany currently imported from South America. Unfortunately, the last Cuban and Dominican trees were cut in the early part of the twentieth century.

Today, most mahogany comes from South America and is sold as "Honduran mahogany" or "genuine mahogany." This species is pleasant to work with and easily available in widths up to 18" on the wholesale market, where it costs about $4 a board foot. You can also readily find wider boards on the specialty market, ranging from $6 to $16 a board foot.

The most important things to look for when buying mahogany are density and interesting grain pattern. Often, the most interesting boards come from common grade lumber, which runs about $1 a board foot less than top grades. Further, in addition to standard patterns, mahogany is also available in distinct figure patterns known as crotch and ribbon stripe. Ribbon stripe appears only in a small percentage of quarter-sawn boards, but is available in the wholesale market. Crotch mahogany sells primarily as veneer.

When buying any rain forest wood, you must take into consideration its environmental impact. In an effort to minimize rain forest destruction, a system of forestry certification has been developed. Independent certifiers have approved certain forests to provide sustainable harvests. Information about this wood is available through the Certified Forest Products Council (see Appendix).

Genuine mahogany.

Black Cherry

Eighteenth century cabinetmakers also favored black cherry. Today, we know cherry best for its deep reddish-brown color, which develops over time. Further, some cherry logs display interesting figured grain patterns. While not quite as stable as walnut or mahogany, cherry is still fairly easy to work and shape. Due to climate and soil content, the best black cherry in the world comes from north central Pennsylvania.

Cherry is readily available at the wholesale level in widths up to about 14" and lengths to 16', and the specialty market offers wider boards, up to 24", and matched cherry flitches. In recent years, as the demand for cherry has risen dramatically, so has the cost. Typical high-grade wholesale cherry now sells for between $4 and $4.50 a board foot. Wide and beautifully figured wood can cost up to $15 a board foot on the specialty market.

Black cherry.

Tiger Maple

This beautiful wood was often used in eighteenth century New England for high quality furniture. Although very rarely used for built-in cabinetry or paneling, its striking appearance, when properly finished, produces an impressive room.

The tiger stripe, or curly figure, appears in both soft and hard maple, though cabinetmakers favor the soft maple for its stability. However, when working with all figured maples, take extra care when planing or molding to prevent grain tear-out.

Tiger maple is available from wholesalers in limited quantities. Because the quality of figure is subjective, it is best to purchase this wood at a retail outlet specializing in figured woods. Buying from a retailer allows you to choose which boards suit your purposes, a choice especially important in figured wood. The price of tiger maple varies depending upon the width of and strength of figure, ranging from $5 to $15 a board foot.

Tiger maple.

Northeast white pine.

Pine

Eighteenth century woodworkers also favored Northeastern White Pine, mostly due to its great availability and to the fact that it was easily sawed and shaped with primitive woodworking tools. They used it in everything from Windsor chair seats to barn siding and wall paneling. Pine's light weight (about 2.5 pounds per board foot), softness, stability, and straight, even grain still make it a pleasure to use.

Throughout New England and the mid-Atlantic states, pine was the first choice for simple furniture, wall paneling, and moldings. Woodworkers most often left this indigenous American softwood unfinished or painted. However, with even the simplest of oil finishes, it takes on a pleasing amber color. Today, cabinetmakers choose it for hand-planed painted reproduction surfaces. Among its other attributes, pine's stability is well-known and appreciated by woodworkers everywhere. However, pine must be kiln dried at a minimum of 150 degrees Fahrenheit to prevent sap bleed-through. If pine is improperly dried, the wood's natural sap can re-liquefy and leak through even painted surfaces.

Today, when purchasing pine for reproduction work, it is important to specify northeastern white pine, which is readily available on the wholesale market. High-end woodworkers most often use grade "C" because it is relatively clear of knots and defects. The wholesale cost of "C" grade white pine is about $2.40 a board foot. However, pine on the wholesale market is only available in widths up to 10". To buy wider pine to make matched wide panels, you must turn to the specialty retail market, where wood up to 25" wide is available (though expensive). Very wide pine can sell for up to $7 a board foot.

Walnut

German settlers throughout Pennsylvania and the Hudson Valley found walnut a favorite, and woodworkers have long prized its beautiful grain and color. Primarily a cabinet wood, walnut was often used for tables and large wardrobes known as kases and schranks. This open-grained wood often has beautiful figure where the trunk forks into two branches. Fine furniture makers seek this crotch wood, as it's called, for its beauty as well as its stability.

The best walnut is a pleasure to work with due to its stability and ease of carving. However, the lower grades can be difficult to use due to the abundance of knots and sapwood. Buying walnut on the wholesale market can be difficult, as the grading laws allow for lots of knots and sapwood in even the top grades. In the retail market, however, you can find incredibly wide and clear boards – as long as you're willing to pay for them. Boards can be obtained from 6" to 36" wide, ranging in price from $3 to $12 a board foot.

Note: Kiln operators often inject steam as they dry walnut, which allows the color of the heartwood to bleed into the sapwood. Although this process allows for better utilization of the sapwood, the final product will look gray and lifeless. Therefore, when shopping for wood to be used in a high-end project, specify "unsteamed" walnut.

Walnut.

Chapter 3
Purchasing Lumber

After completing the project design and making the material choices, next purchase the lumber for the project. Buying lumber for fine cabinetry is not as simple as it might appear, and picking out just the right wood for any given project does not lend itself to one-stop shopping. Rather, take an eclectic approach, and at least explore the three main venues for buying lumber.

Wholesale Suppliers

Wholesale suppliers are scattered throughout the country and deal in kiln-dried graded lumber. To buy lumber from them, you must have a basic knowledge of the National Hardwood Lumber Association (NHLA) grading rules by which they operate (available in the NHLA handbook). These rules are all based on the percentage of usable wood in a given board. The acceptable percentages vary from species to species: for example, a walnut board can have more knots than a poplar board and still qualify for the same grade.

The advantage of buying from a wholesale distributor is consistency in wood quality and moisture content. Also, wholesale suppliers generally deliver the wood right to your door within a few days.

One disadvantage of this buying method is that they generally have a minimum 500 board foot purchase (but which is not a problem if you're building an entire room). Also, you cannot pick and match individual boards. This doesn't matter particularly in a paint grade room, but makes a big difference in a finished hardwood room. According to my experience, wholesale sales representatives are very knowledgeable and eager to help with any questions. For a more in-depth understanding of how lumber is processed, give one a call and schedule a tour of the facility. (See appendix for wholesale distribution.)

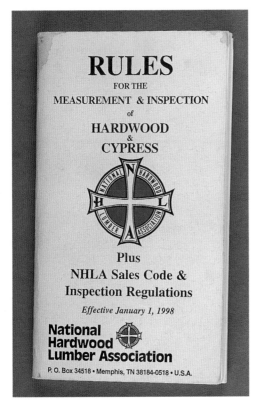

NHLA handbook.

Grading lumber. *Photo courtesy of TBM Hardwoods.*

A wholesale lumber yard. *Photo courtesy of TBM Hardwoods.*

Air-drying lumber. *Photo courtesy of TBM Hardwoods.*

Loading a commercial kiln. *Photo courtesy of TBM Hardwoods.*

Electronic wood tally. *Photo courtesy of TBM Hardwoods.*

Retail and Specialty Lumber Dealers

Buying lumber retail is the most fun way to purchase lumber for any project. A good retail supplier has thousands of BF of lumber to choose from. Many keep some of their lumber in a flitch (all the boards which came out of a single log together), which makes matching panels possible. Some suppliers have boards from 20 - 50" wide.

When buying wood, be certain to bring along a moisture meter to check the moisture content. (See appendix.) A moisture content of 5-10 percent is allowable for most projects. One problem that many retail sales outlets have is that their lumber often sits in unheated buildings for a long time, picking up excess moisture. (12-14 percent in many parts of the country.) If this happens to a particular flitch you want to buy, simply sticker it in a heated room for a couple of weeks until it reads 6-8 percent moisture content.

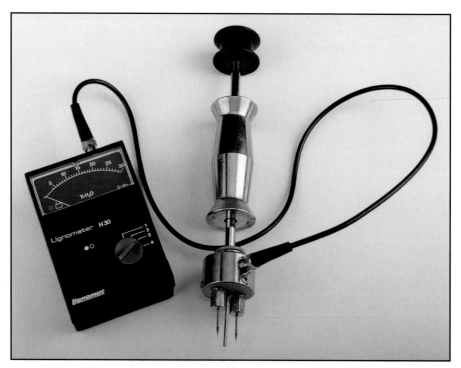

Moisture meter.

When selecting a retail supplier, make sure the supplier will allow you to pick out your own lumber. The only down side to retail and specialty lumber is cost. The wide, beautiful boards they can offer are difficult to find, so retailers charge accordingly. (See appendix for retail lumber.)

Retail lumber selection. *Photo courtesy of Hearne Hardwoods.*

Splitting large log before sawing. *Photo courtesy of Hearne Hardwoods.*

Buying Logs

Whether buying logs from a local sawmill or out of your neighbor's back yard, the process of selection, sawing, and drying is about the same. Buying logs allows you to obtain wide flitch-matched lumber at a much lower cost than by any other method. However, any mistakes made along the way can be very costly and result in low quality lumber.

The following information is meant to be an overview of the process rather than a complete detailed guide. There is simply too much information (much of it experiential in nature) to be contained in one chapter of a book on cabinetry.

A typical log yard. *Photo courtesy of Hearne Hardwoods.*

Step one: log selection

The best place to start looking for logs is at small saw mills. These mills are scattered throughout the country, usually in rural areas. Small mills usually have a good supply of logs native to their area, but rarely stock logs from more than 100 miles away.

Buying logs from a mill operator is similar to horse trading, as logs do not come with price tags and vary dramatically in price from sawmill to sawmill. Especially large or unusual logs are priced even more subjectively than normal saw logs.

Correctly judging the quality of a log takes experience, and even professional log buyers make mistakes. However, there are several things to keep in mind when buying a log:

How long ago the log was cut. Many species deteriorate due to internal fungal growth, particularly during warm months. Look for evidence of staining.

Rate of growth. Check the annual rings for growth rate. Usually the slower growing trees are more valuable.

Knot defects. It's easy to see where branches have been sawed off of a log. Check further to find any irregularities in the log where a branch might have broken off years ago and healed over. These imperfections indicate where there will be knots, usually going though the tree all the way to the heart.

Checks and shake. A check going all the way across the log on both ends is normal and no reason for concern. However, any cracks running parallel to the annual rings often mean wind-shake. Pass over such logs.

Off-center heart. Tree hearts are often slightly off-center. This condition makes very little difference to the cabinetmaker, but when the heart is dramatically off-center, it often means the tree grew on an angle, and any wood sawed from this log will be under too much tension for cabinetry.

A portable saw mill. *Photo courtesy of Albert Bartell.*

Bandsaw mill cutting. *Photo courtesy of Albert Bartell.*

Sawing Logs

Often, logs are sawed where you purchase them, i.e. at the sawmill. When having a local sawmill cut your lumber, make sure they understand exactly how you want it cut. Or, have a portable sawmill brought to the logs. Portable mill operators are accustomed to custom sawing and usually charge by the board foot or by the hour. They often have the capability to cut up to 28" wide and hitting a nail or other metal with a band saw blade costs a minimal amount compared to hitting it with a circular mill.

Drying Lumber

Drying lumber would seem to be as simple a process as evaporation itself, but more lumber is ruined during the drying process than at any other time. If you are planning to take the lumber to the kiln when still green, do it within one day after sawing in the summer months and 3 days after sawing in the colder months. If you choose to dry the lumber yourself, you will need a very flat surface on which to sticker the wood. This surface can be concrete or pressure treated 2 x 4s placed on edge directly on the ground on 18" centers. Most importantly, the surface must be flat.

Then place the lumber on the 2 x 4s with about 1" air space between the boards. Space each consecutive layer with stickers (0.75" x 0.75" boards 4' long) until the stack is complete.

Leave the unit outside during the air-drying process, but cover the top with tin or a similar material to protect from rain and sun. (Do not cover the sides.) The wood should then air dry in this stack until it reaches about 20 percent moisture content. This usually takes 4-6 months, depending on drying conditions. If the wood you are drying is susceptible to fungus, as maple and many other white woods are, you should replace the stickers with fresh dry ones after about four weeks.

Stickering a unit of cherry. *Photo courtesy of Good Hope Hardwoods.*

A properly stickered unit. *Photo courtesy of Good Hope Hardwoods.*

36

When the wood has dried to about 20 percent, you can then sticker it inside a heated space to finish drying, or take it to a commercial kiln for finishing. Woodworkers generally agree that this system of air-drying first, then finishing in a kiln produces the best lumber.

Loading wood inside kiln. *Photo courtesy of Good Hope Hardwoods.*

Chapter Four
Cabinetry and Paneling

History

No one knows for sure when the first "floating panel" was used, but ancient Egyptian woodworkers as far back as 3000 BC understood the need to allow for wood's seasonal expansion and contraction. In the early eighteenth century, people used wall panels primarily as protection from cold weather. This primitive paneling was usually ship-lapped or feather-edged vertical boards spanning from floor to ceiling. Later, this simple means of sealing a home from drafts evolved into a decorative form we now call Georgian style paneling.

Overview

Now that the drawings are complete and the wood has been chosen and purchased, you are ready to begin constructing the various elements of the room. In this chapter, we will discuss the making of cabinets, cabinet doors, and wall paneling. The photos accompanying this section use a cabinet with plywood carcass parts, beaded face frame openings, and raised panel doors for demonstration.

Carcass Construction

Make an organized cut list based on the job's drawings.

Beading the inside edges of the face frames on a router table.

Before cutting any wood, make a complete cut list of all the face frame parts. This cut list should indicate the exact dimensions of each piece of wood in the project. Having a complete and exact cut list on hand will help you pick out the rough lumber needed for the job. After milling the rough lumber to its final size, mark each individual piece where beads, mortises, and grooves will be located. Next, run the bead on the inside edges of the face frame parts. Then, groove the face frame parts on the back to accept plywood carcass sides.

Grooving the backs of the face frames to accept plywood.

Because our example face frame has beaded openings, the parts require a special joint. Cut the beads of the male part in a chop saw at a 45 degree angle, and cut the female mitres on a table saw also set to a 45 degree angle. Then, clean up the waste with a chisel. These two parts are now ready to be mortised. You may cut a mortise several ways, including with hollow chisel mortisers, hollow chisel setups on drill presses, slot mortisers (shown here), routers, and the time-honored chisel and mallet.

The waste is then pared out with a chisel.

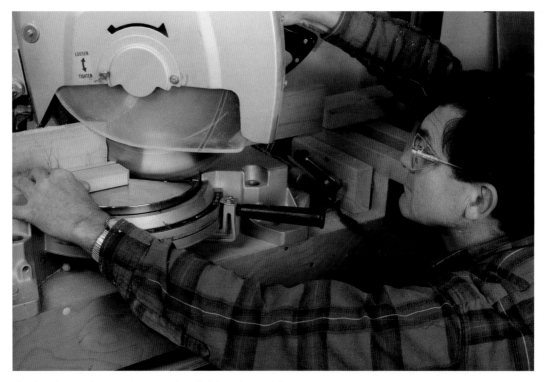

Cutting the 45 degree mitre on a beaded face frame joint.

Beaded face frame joint, with floating tenon ready to assemble.

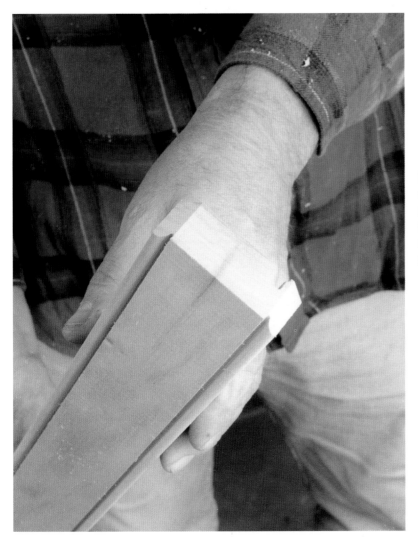

In this example, we cut two mortises, one in each piece. We then join the two pieces by a floating tenon. Alternatively, you can cut the male piece with an integral tenon. We use the floating tenons because the setups are more efficient, and their strength is equal to the hard tenons'.

Face frame part with beads mitred.

Beads mitred in stile can be cut with a mitre trimmer or on a table saw.

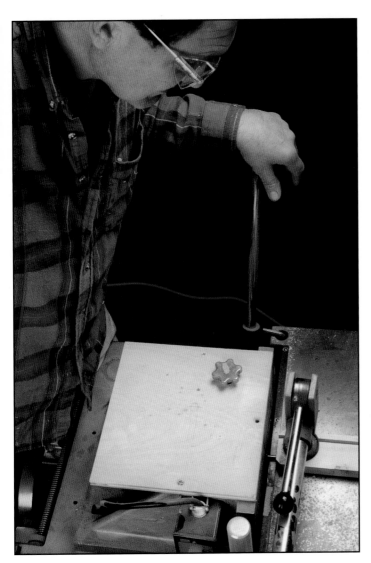

Mortising a face frame part with a slot mortiser. This operation can also be done with a router jig or chisel and mallet.

Mortised face frame part.

When you have made all of the joints, dry fit all of the parts together. Lay them out on a workbench, make sure the tenons fit, and push the parts together. Clamping the parts should make your face frame straight and square. Check that the diagonals are equal by measuring from corner to corner. If the face frame passes the dry fit test, add glue to the joints and clamp them together.

Face frame parts should slide together easily without being too loose.

Clamp parts together dry to check that they're square, the correct size, and so on.

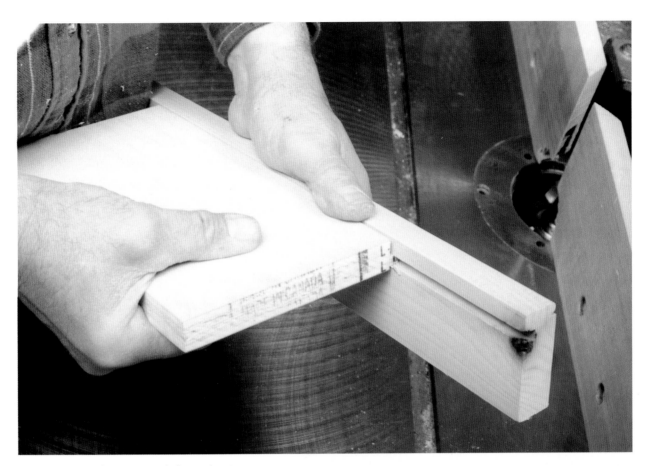

Fitting a tongued test piece of plywood to its groove.

While the face frames dry, you can make the plywood parts. First, put together a cut list of all plywood parts, making sure to allow extra length for dados and extra width for tongues. You can then simply cut the plywood sheets to the proper size. Tongue each piece on the shaper, first running a test piece to check for proper alignment. Save this test piece for later use when marking the dados.

Using a test piece of plywood to mark the dados.

To mark the dados, place each piece in its own face frame groove, place the test piece in the intersecting groove, and make a pencil mark. You can cut the dados on a panel router (shown here) or with a hand-held router and straightedge. When you have dadoed all of the pieces, assemble the whole box without glue and make sure that everything fits. When you have checked everything, disassemble the pieces and prepare a workstation for gluing, having plenty of glue and clamps on hand.

To glue each piece, starting with cabinet sides, add a small bead of glue to the inside edge of the grooves, press the pieces into place, and clamp. The test piece from earlier can help you quickly check the alignment of the dados as you glue. When you have glued all of the plywood parts, screw them together through the dado joints. When the glue is dry, make up appropriately sized pieces of .5" plywood for the backs, and attach them to the carcass sides with screws.

Cutting a dado with a panel router. This can also be done on a table saw or with a router and fence.

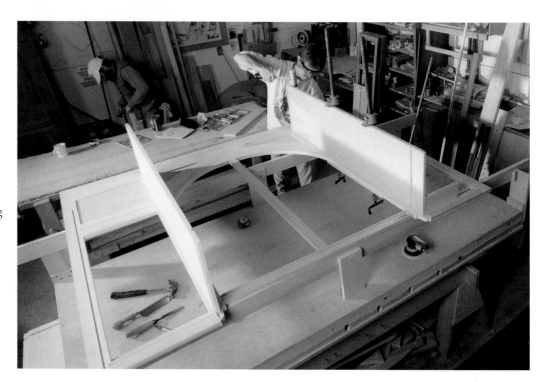

Dry assemble all parts, starting with the side pieces . . .

. . . and finishing with the top and bottom.

Apply a small bead of glue to the insides of the groove.

After applying glue to the groove, slide each piece into place . . .

. . . and clamp firmly.

Plywood pieces are then screwed together through the dados.

Panel Construction

After consulting your drawings and making a complete cut list, cut the rails and stiles to size. Cope and stick them on the shaper, using the same methods as in cabinet door construction. For larger panels, put tenons between each cope and stick joint for extra strength. Dry fit the frames, and measure each opening to find the exact size of each raised panel. Rough cut the ramps on the table saw, and run the panels through the shaper. Check that the panels fit and that the panel tongues slide into their grooves easily without being too loose. Disassemble the parts and prepare for gluing.

Gluing up large panels can be problematic, but with proper attention to efficiency, you can overcome these difficulties. To start, glue only the center dividing stile or stiles. Clamp these pieces to their rails, and check for alignment by sighting down each stile and measuring the panel openings. Slide in each panel, and check that the openings are still square. Next, apply glue to the outer stiles and their corresponding rails, and push the parts together. Check that the panels are centered in their places. Do this now, because accidental glue squeeze-out inside the joints can "freeze" the joints in place. Clamp the panels, and do a final check for alignment and squareness.

Cabinet Doors

Measure all door openings and make a complete parts list. First, make the rails and stiles for doors, dry fit them, and take the panel sizes from the assembled frames. In this chapter, we use a three-piece cope and stick shaper cutter set for door construction. This method, although not strictly the joinery used in the eighteenth century, produces a look almost indistinguishable from the eighteenth century doors at a fraction of the time.

To construct the door, first cope the door rails. Cope moldings interlock with the stick moldings and create an interlocking glue joint. Cut the rails a little bit longer than the final size to allow for waste from the shaper; 0.0625" is usual. You can make this cut most easily with a shaper with a sliding table, but a carriage made to slide in a mitre gauge slot works quite well if the sliding table is unavailable.

After coping the door rails, run the sticking on all door parts. Cut the door parts about 0.0625" wider than usual to allow for two passes of 0.03125" on the shaper. We use two passes because often the first cut is deep and tears out some of the grain; the second cut cleans this up. Set up the shaper to match the height of the cope moldings, and run a test piece. Check that the fit is exact, and then run the actual stock. A bit of extra care in these setups can save a lot of sanding later.

Dry assemble the rails and stiles in each door. Check that the size of the door is correct for each opening, and measure for panels, allowing for the panels' tongues. Cut panels to size and rough cut the ramps on the table saw. (For more detail, see Chapter Five, "Passage Doors.") Raise the panels in the shaper and check the panels' fit in their grooves. Sand all parts, and prepare for assembly. To make the sanding of panel ramps easier, we recommend a small pneumatic random orbit sander with a 1" disc. These look like small angle grinders and can be purchased at some auto parts stores and a few hardware specialty stores.

Before gluing, dry fit each door to check that all parts are correctly made and assembled. Next, apply glue to each joint and clamp. Check that the panel is centered in its opening and that the door is square. Let the glue dry for 1 to 2 hours, and remove the clamps.

To hang the doors, rough fit them in their openings so that they go in easily, but are a little tight. Designate the hinge side, and shave off an additional 0.03125" and back bevel it by about 5 degrees. A jointer works very well for this. Mark the hinge mortises and cut them out. Attach the hinges and mark their placement on the cabinet by gently sliding the door into its opening. Make marks simply by drawing a pencil line where the hinges fall. If you have some room up and down, shim the bottom slightly. Next, rout out the cabinet hinge mortises, and attach the door. Check the fit, and pare away the door where necessary. Back bevel the entire door to finish.

Unassembled cope and stick joint. The stick is essentially a groove made to accept the panel tongues with a molding on one or both sides. In this case, the molding is an ovolo. Cut the cope on the ends of the rails, which fits the stick exactly. A .625" deep tongue and groove provides sufficient gluing surface, so you don't need an additional tenon.

Cutting the cope on a shaper with a sliding table.

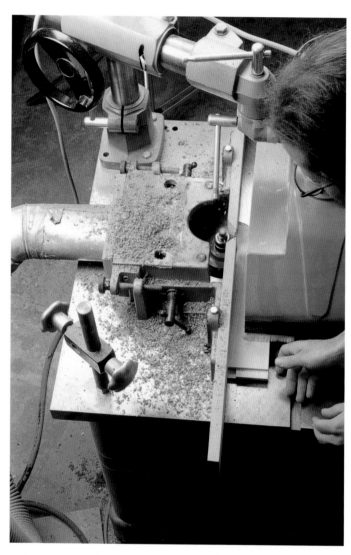

Raising a panel on a shaper.

Running the stick molding on a shaper. The power feeder (shown here) is not necessary, but does make the cuts more consistently and safely.

Door parts ready to assemble.

Applying glue. Too much glue can make the door appear sloppy and can freeze panels in place; however, take care to apply a sufficient amount. Apply glue to both pieces.

Attach one rail and slide panel into place.

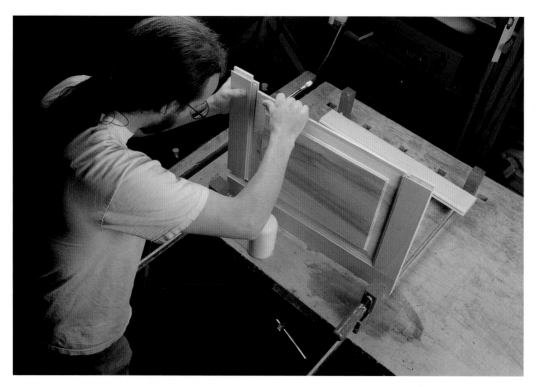

After properly seating the panel, attach the other rail.

Apply glue to the rail ends and stile . . .

. . . and clamp door together. The K-body clamps are less apt to cause the door to buckle upward, but almost any clamp will do.

Make sure that the panel is now centered in its opening, so that if any glue accidentally seeps into the corners, the panel is properly placed.

Check that the door is square by measuring the diagonals. If the two diagonal measurements are equal, the door is square.

Mark the hinge mortise placement on the door. Usually, the hinges are placed in-line with the shoulder of the stick molding.

Once the hinge mortises are marked, score the ends with a chisel. This makes the routing operation cleaner.

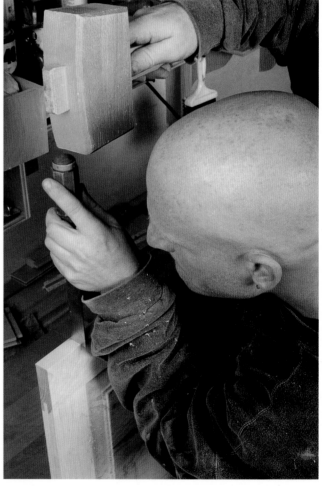

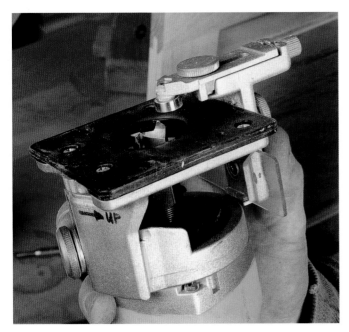

A laminate trimmer with an edge guide, like this one, makes routing the hinge mortises much easier.

Pare away the remaining waste with a chisel.

Routing a hinge mortise with a laminate trimmer.

Check that the hinge fits, and screw it on.

Two dimes work well for getting a reveal for the door placement.

Place the door, with hinges attached, on top of the dimes, and mark where the hinges lay.

Extend the lines with a combination square, and route the hinge mortises.

Shelves

When the cabinetry carcasses are complete and the doors are hung, fit the boxes with shelves. In the eighteenth century, most shelves were fixed in one position. Today, however, many customers request adjustable shelving. To accomplish this, we borrow a method often found in antique secretary desks – ratchet strips. This system of ratchet strips requires more time to construct than other more common methods (such as drilling holes on 2" centers and inserting brass pins for shelf support), but it adds interesting detail and works very well.

To make ratchet strips, first measure the interior of the carcass for height and make the blank pieces accordingly. Blanks should be as thick as the ratchet strip is wide, and wide enough to cut into at least four strips per blank, as you need four identical pieces to make the shelves level. Mark the blanks to show where each ratchet falls.

Cut the ratchets using a radial arm saw with a shop-built jig and a pair of custom-cut saw blades. Each blade is cut with a 60-degree bevel, one right-handed and the other left-handed. The jig should also have some hold-down clamps installed to keep the blanks from shifting during the cutting operation, especially when making the angled cut.

Saw the angled cuts first, making sure that the points of the cuts fall below the marks on the blanks. Make these cuts with the saw tilted to 60 degrees. When you've finished all of the angled cuts, set the saw to vertical and make the vertical cuts. The tips of the cuts should meet perfectly. When you've cut all of the blanks, saw the blanks into strips. Sand the strips to thickness, and install them into the cabinets with pin nails.

Make hangers as thick as the ratchet strips, cut on the ends with angles to match the ratchet strips.

Make shelves as you would normal shelves, but cut out the corners to accommodate the ratchet strips. A jig made to fit the table saw that holds the shelf vertical works well. Set it up to cut out the corners with a dado stack. Clean up the corners with a chisel to finish.

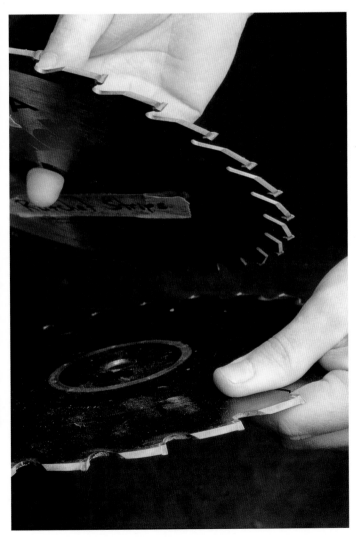

To cut ratchet strips on a radial arm saw, you must custom grind two blades with 60 degree angles.

Make a mark on the saw's fence that corresponds with the tip of the blade. Then, align the marks on the ratchet strip boards with the blade mark.

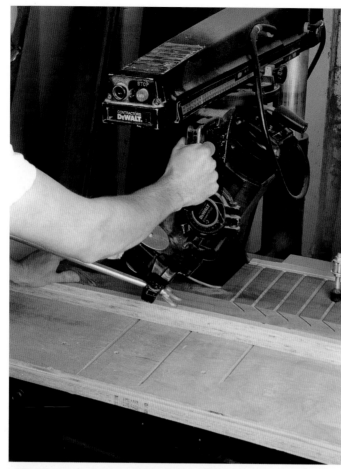

Ratchet strips with a radial arm saw, set at 60 degrees.

Below:
Doing the second cut with the radial arm saw set at 90 degrees.

Ratchet strips glued and nailed into place with a pin nailer.

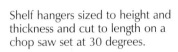

Shelf hangers sized to height and thickness and cut to length on a chop saw set at 30 degrees.

Shelf hangers should fit securely with a small amount of side-to-side play.

Beaded edges can be cut with either a shaper or router.

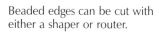

Beaded shelf. This molding was cut with a custom-made 0.1875" bead shaper cutter.

Cutting notches with a table saw. The table saw has a 0.75" dado stack mounted on the arbor, and this shop-built jig keeps the shelf in position.

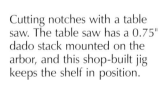

Detail of notched shelf. Once the cut is made, the notches should be cleaned up with a chisel to even out the dado marks.

Once the shelves are made, check that they fit into the cabinet correctly. They should be slightly loose to allow for seasonal change in the wood.

Detail of installed shelf and ratchet strips.

Chapter Five
Passage Doors

Early in the eighteenth century, doors were usually very simple, often a few ship-lapped or feather-edged boards with battens attached by clinched nails. The hinges were usually iron Hs or H & Ls, though butt hinges appeared a little later in the eighteenth century. As the merchant class grew and farms became more prosperous, doors (as well as other millwork) evolved as important symbols of wealth and prestige.

Because individual tastes and regional differences did not follow a strict timeline, precisely dating a door based on its design poses some difficulty. However, a few general rules exist. For example, almost all eighteenth century passage doors are thin by today's standards – usually only 0.9375" to 1.125" thick. Most of these doors had sticking and raised panels on only one side. Also, eighteenth century houses' downstairs doors were often of higher styles than other, less visible house doors.

Today, manufacturers produce doors by the fastest possible methods and from a variety of materials. These doors often prove low-quality and almost always result in sub-standard appearance when used in conjunction with a custom paneled room. For this reason, we have chosen to illustrate how to faithfully reproduce a true eighteenth century door.

First, consult the drawings and make a cut list. Mill all frame parts to length and width. Then, run the sticking along the inside edges of the frame parts. You may purchase special cutters in matched cope and stick sets, as shown. Some, like the one pictured here, utilize a rub collar that does not cut the back shoulder. When using a shaper with rub collar cutters, set the cutter height and make sure the fences and rub collar are on the same vertical plane. Other cutter sets, more convenient and expensive, include a cutter for the back shoulder. A router or hand plane can accomplish this same task.

With a piece of scrap cope molding, mark where the mortises in each stile and rail will appear. Make light cuts on the stick molding, and pare out the waste with a chisel. When using a hollow chisel mortiser, as we do in this example, cut a series of independent squares, and then cut the spaces in between. This prevents the bit from "wandering" while cutting. If the mortise is too deep for the bit, square down to the other edge and cut from the opposite side. You can also cut these mortises with a router or chisel.

Half-inch sticking cutter with rub collar.

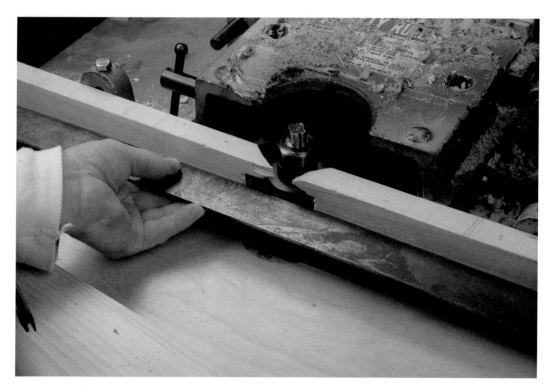

Using a straight edge to align the fences with the rub collar.

Running a piece through the shaper.

A properly shaped piece.

Using a scrap piece of cope molding to mark the position of a mortise.

Cutting the stick molding.

Starting with the ends, . . .

. . . pare out the waste with a chisel, . . .

. . . and finish with the corners.

Making a series of separate square cuts.

Removing the remaining wood in the mortise.

Squaring down from front to back.

Mortising the back side.

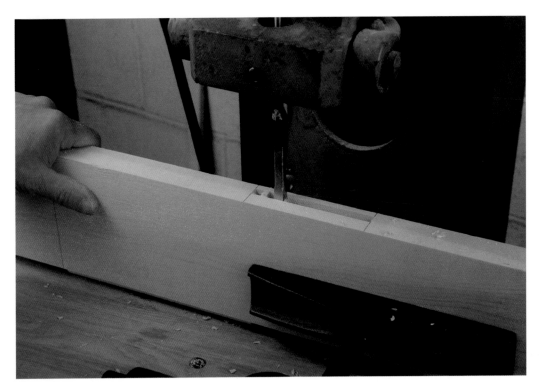

Completing the mortise.

Setting the table saw blade height with a square.

To cut the tenon, mark all shoulders – the first parts to be cut – with a square on a test piece, set the table saw blade so that it cuts to the depth of the tenons. Then, when the setup is correct, run the actual pieces. Set up the blade and fence for the other side of the tenon, again cutting a test piece before running the actual stock. The pieces will then have offset shoulders. Remove the rest of the waste with either a dado stack and mitre gauge, or a vertical tenoning jig. Then mark and cut out the width of the tenons. Make sure that the tenons are slightly narrower than the mortises to allow wedges to be driven in later. In this example, we coped the top, bottom, and lock rails in the sticking only. Cut the corners out roughly on the band saw, and make the final cut with a .5" radius incannel gouge. Check the pieces for fit. Sometimes, back cutting the cope just a bit helps the joint to come together. On this door, we coped the dividing stiles as usual on the shaper.

Running a test piece. Note: when using a mitre gauge, clamp a standoff block to the table saw fence.

Running the tenon shoulder on the back of the actual piece.

Adjusting the fence for the front side tenon shoulder.

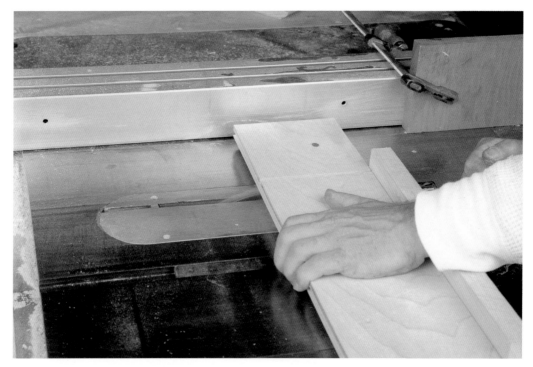

Running the second tenon shoulder.

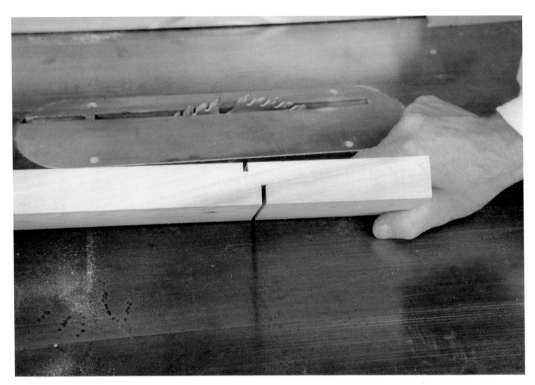

An example of properly cut tenon shoulders.

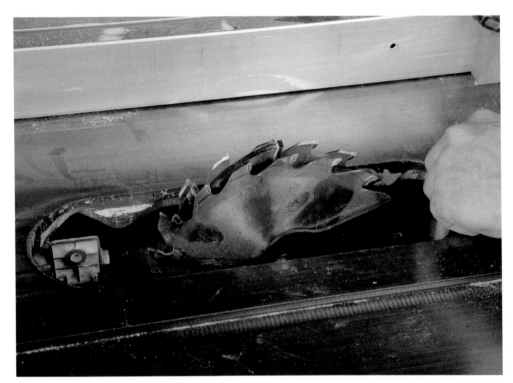

Installing a dado stack on the table saw.

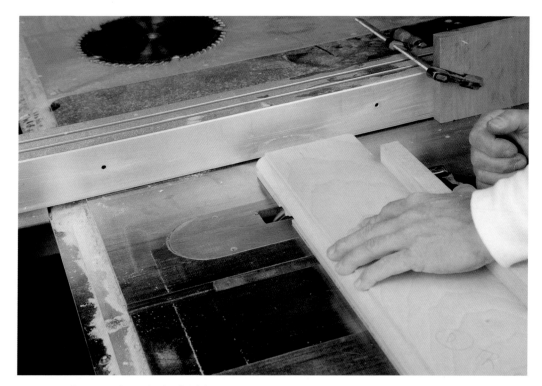

Removing the waste from the back of the tenon.

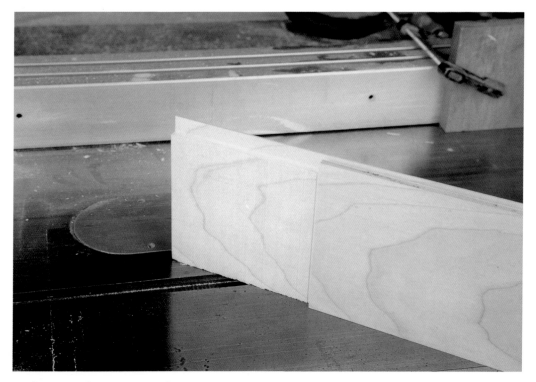

Work piece with waste removed.

Removing the waste from the front side of the tenon.

Measuring the distance for the side of the tenon.

Cutting the side of the tenon with a band saw.

Rough cutting the cope with a band saw.

Tenons before and after band saw work is completed.

When you've made all the frame parts, assemble and make certain that the frame is square. Check that all of the joints are tight, measure for panel sizes, and cut the panel stock. Before running the actual profile on the shaper, make a rough cut on a table saw first, especially when using a shaper with a 1" or smaller arbor. Make a jig with the appropriate angle and clamp it to the table saw fence. After setting the blade height and fence positions, make four cuts, starting with the end grain cut. When you've rough cut all of the panels, raise them on the shaper, starting with the end grain cut to reduce tear-out. Sand the ramps and, if desired, hand plane the panels and frame parts.

Assemble all of the parts, and clamp. You may gently tap parts that are out of alignment into place with a hammer and a piece of soft wood. After clamping and checking that the door is square, drive wedges into the spaces left on either side of the tenons. Lastly, mark the pinholes and drill them out. Make some square peg stock, slightly round their corners, and drive them into the door.

Note that this method requires no glue: The doors hold themselves together. The pins and wedges will keep the door intact for hundreds of years.

Cutting the cope with an incannel gouge.

Completed cope.

Test fitting a mortise and tenon joint (close-up view).

Door assembled to check for proper fit.

Checking that joints are tight.

Measuring the openings for panel sizes.

Table saw setup to rough cut panel ramps.

Using a previously made panel to set up table saw for rough cutting the panel ramps. (View from face.)

Using a previously made panel to set up table saw for rough cutting the panel ramps. (View from panel edge.)

Rough cutting the panel ends.

End of board as rough cut.

The side ramps of the panel are cut second.

Panel with rough cutting operation complete.

Running a panel through a shaper with a raised panel cutter.

Cutting the end grain of a panel tears out the side of the panel. Removed this after you have run the side ramps.

Raising the side ramps of the panel.

Raised panel with shaper operations complete.

Hand planing the face of a raised panel.

Assembling a door, placing rails into one stile first.

Sliding in the panels.

Aligning dividing stiles.

Dividing stiles should line up.

The remaining panels are then added.

Installing remaining stile.

Assembled door.

Place one clamp on an end.

Clamping the dividing stiles.

If dividing stiles do not line up, gently tap them into place with a hammer and a block of soft wood.

Mortise and tenon joints should have a small space towards one edge to allow for wedges.

Corner pins and tenon wedges.

Tapping wedges into mortise and tenon joints to make them tight.

Making a diagonal line from the cope corner to the door corner.

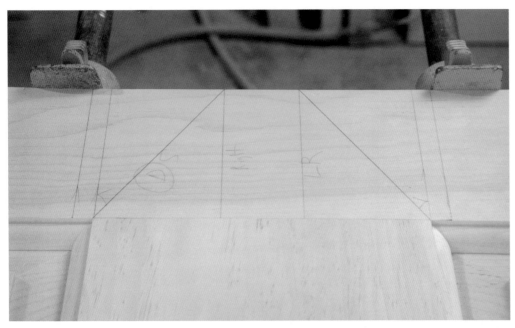

Stile with pin marks.

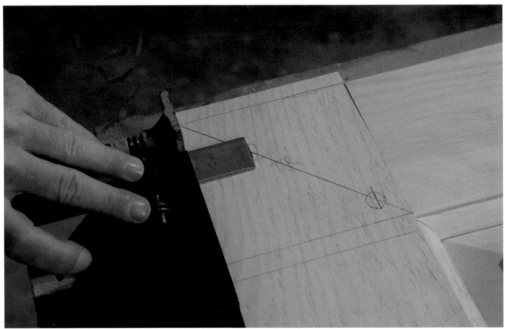

Using a combination square
to mark pin placement.

Drilling holes for pins.

A completed door
with faux finish.

Chapter Six
Moldings

Moldings create smooth transitions between architectural elements. For example, a cornice detail can visually connect a paneled wall to the ceiling. However, molding details like pilaster strips can also exist as their own architectural elements, because they appear to be a structural support, holding up a ceiling or header.

Detail of pilaster and cornice.

When creating an eighteenth century room, using eighteenth century molding patterns, rather than twentieth century lumber yard patterns, is essential. Designing your own molding patterns is unnecessary. Instead, collect them by visiting old houses, or obtaining books with good pictures or line drawings of molding details. Be careful when following molding patterns that the house is authentic, and that you are not taking patterns from someone else's restoration work. When doing your research, also make sure that you know when the room was built, as it may vary considerably from when the house was built. In this manner, you should quickly begin to notice differences between moldings from various time periods and geographical locations.

Many assume that it takes dozens of cutters to make an eighteenth century room. However, with a little ingenuity, you can create a wide variety of moldings using only a few relatively inexpensive cutter patterns. We will show some of the many patterns you can create using just ten cutters.

To produce the moldings you have chosen, first dimension and surface all four sides of your lumber. Flattening the wood on a jointer, and then plane it. Finally, joint and rip the edges. Next, decide which equipment is best suited to make the variety of moldings needed to put together a room. The five most common tools used in making moldings are the shaper, molder, router, table saw, and hand plane.

0.75" ogee cutter for 0.75" shaper.

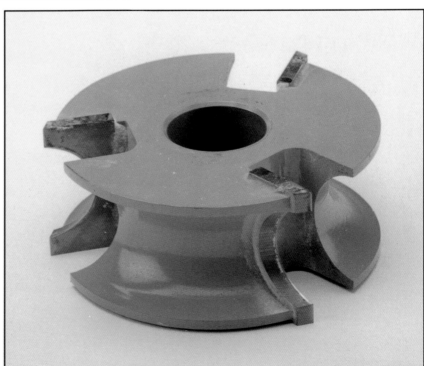

0.75" bead cutter for 0.75" shaper.

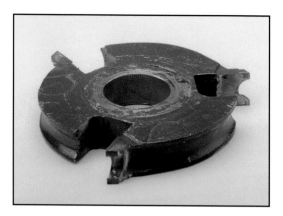

0.25" bead cutter for 0.75" shaper. Due to the difficulty in finding cutters with 0.0625" flats, we had one custom made.

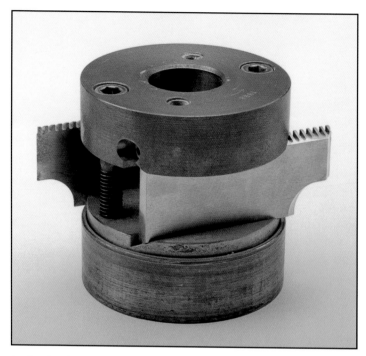

Lock edge shaper cutter for 1.25" shaper. Ovolo cutter is custom made.

0.375" bead cutter for 0.75" shaper. This cutter also had been cut so that the flats are only 0.0625".

Cope and stick set, and raised panel cutter. These cutters in combination make cabinet-sized doors with 0.25" quarter round sticking with square shouldered raised panels.

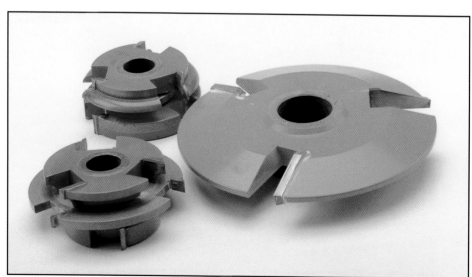

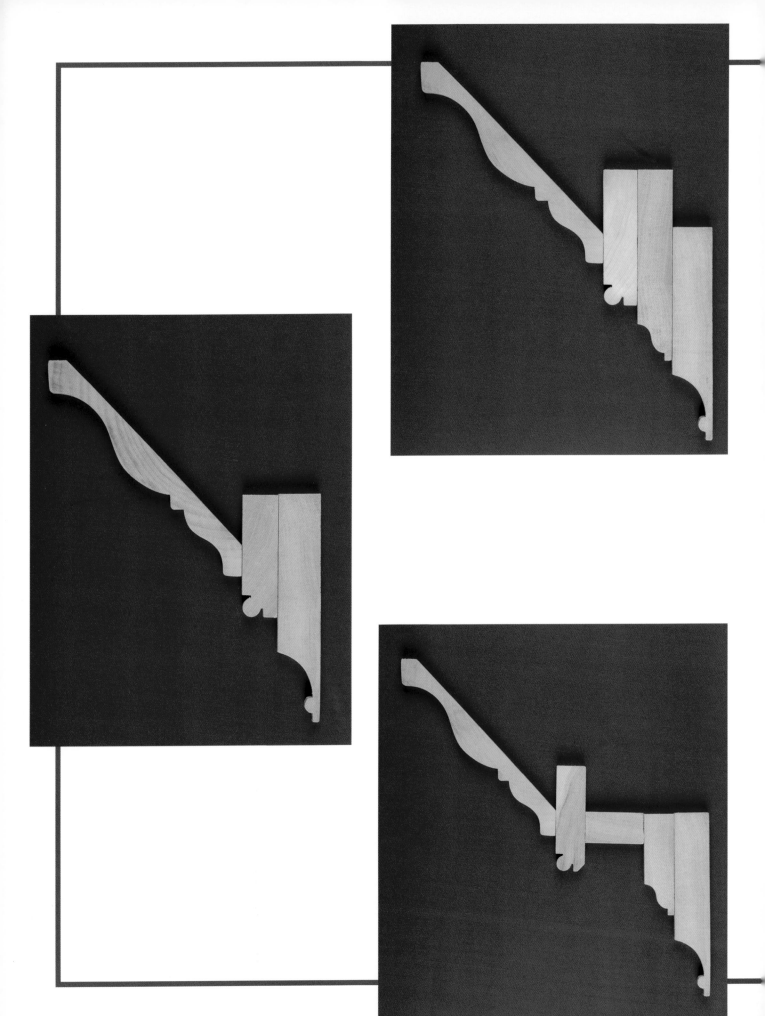

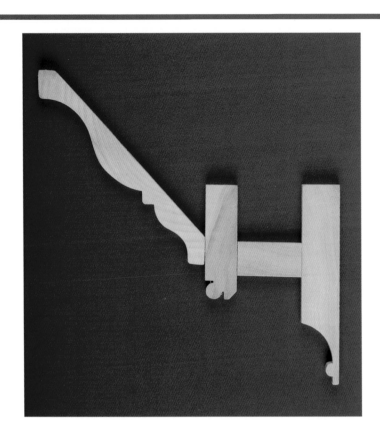

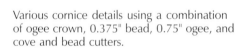

Various cornice details using a combination of ogee crown, 0.375" bead, 0.75" ogee, and cove and bead cutters.

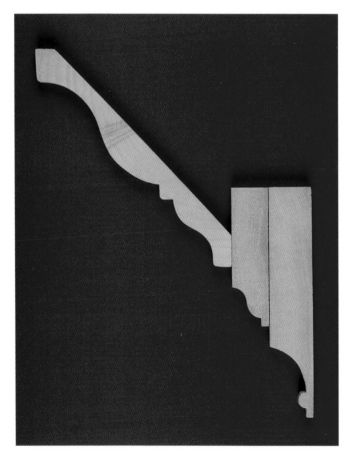

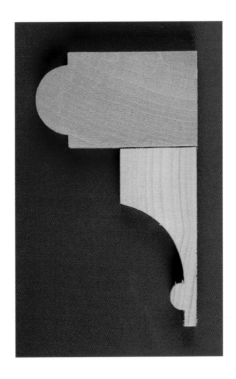

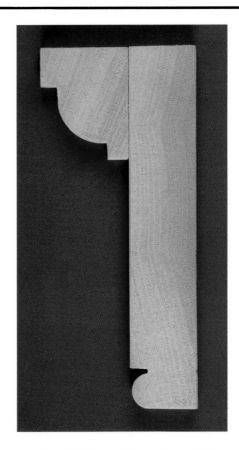

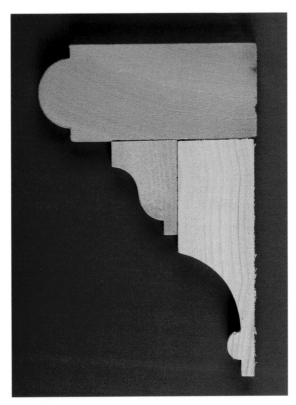

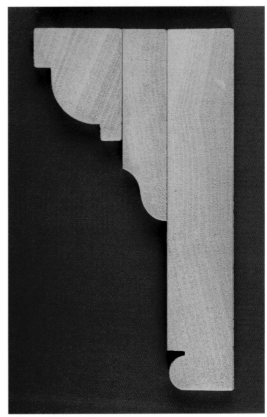

Two chair-rail or wainscot cap details, using a combination of 0.75" bead, cove and bead, and 0.75" ogee cutters.

Two case molding and backboard details using various combinations of ogee, 0.375" bead, and 0.75" ogee cutters.

Detail of baseboard with ovolo shoe molding and base cap using base cap cutter and quarter round cutter from the cope and stick set.

Detail showing the interlocking fit of cope and stick moldings.

Flattening the wood on a jointer.

Detail of quarter round stick moldings and square-shouldered raised panels. Note that the gap between the edge of the panel and the stile will allow for seasonal expansion of the panel.

Planing wood to proper thickness.

Edge jointing.

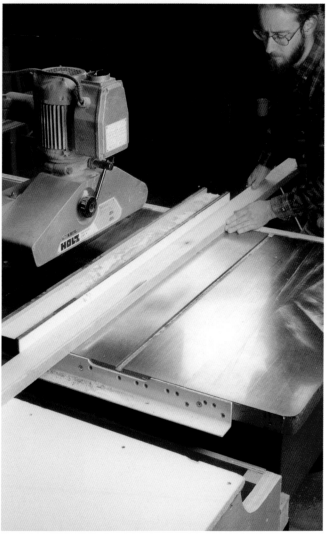

Running wood through a table saw. Note: guard has been removed for clarity.

Shapers

Shapers are perhaps the most versatile tools available. They come in different sizes based on the diameter of the spindle, most commonly 0.5", 0.75", 1", and 1.25". Determine the size of the spindle by the size of the molding you want to cut. For instance, a 0.5" spindle can easily cut small moldings such as beads. However, for heavier cuts, like raised panels and crown molding, a 1.25" spindle works best. For the amateur woodworker and small cabinet shop, the most popular shaper is the 0.75" because of its versatility and the lower cost of both the shaper and cutters. For even greater flexibility, many shapers are available with interchangeable spindles from 0.5" to 1.25".

To set up a shaper, you need a straight edge, a ruler with end gradations, and, of course, a pair of good wrenches to tighten down the cutter. Use the straight edge to align the fences, and the ruler to set cut depth and cutter height. Before doing *anything* on a shaper, or any other tool for that matter, be sure to turn off the power at its source.

The common types of cutters for the shapers are as shown (starting clockwise from the upper left-hand corner): corrugated molding cutter, three wing fixed cutter, lock edge molding head, lock edge molding head with bearing, three wing sticking set, three wing coping set, and two wing raised panel cutter. The last cutter shows a separate bearing on top, which can be used to create curved panels and moldings.

Corrugated backed molding head cutters are fairly easy to set up and adjust and are excellent for making large moldings. However, the relatively high initial cost of the head and custom cutters can hinder their purchase.

Three wing cutters are the most commonly used cutters on a shaper because of ease of setup and relatively low cost. These cutters are available in hundreds of patterns, and often are carbide tipped, which reduces sharpening costs.

Lock edge cutters are often used when custom-ground cutters are required. Many shops grind their own cutters; you can also order them custom-ground from high-speed steel. However, lock edge cutters take an extra few minutes to adjust and set up.

You can purchase cope and stick three wing cutters in matched sets in a variety of shapes for making doors and paneling frames.

Two wing cutters are often available as custom cutters and can be carbide tipped and ground to a custom shape at a somewhat lower cost than a three wing cutter.

Use a straightedge to align the shaper fences.

End gradations on a 6" ruler make it easy to set the cut's depth.

The end gradations also help with setting the cutter height.

Various shaper cutters.

Molders

Molders are usually priced beyond the means of the small cabinet shop, but several companies make smaller machines that, although somewhat slower than the full-sized tools, still produce high-quality moldings. These best cut profiles on large flat surfaces, such as crown moldings, rather than on edge moldings, which are more easily done on a shaper. Williams & Hussey molders work with an infinite variety of cutter shapes to produce high-quality moldings. Knife placement and setup take about ten minutes, but the stock must be squared before it is run through the machine. For making crown molding, you may purchase an additional back knife, or run the workpiece through the table saw, creating any angle of crown desired.

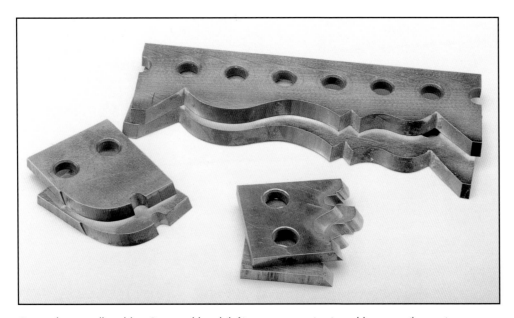

Cutters for a small molder. Cove and bead (left), ogee crown (top), and base cap (bottom).

Williams & Hussey molder/planer. The wooden piece in the foreground is for attaching auxiliary wooden tables.

Various cutters custom made for Williams & Hussey molder/planer. Cost: about $40 per inch of width.

Bolting knives onto the cutter head.

Placing wood in its proper position, and screwing fences down to the auxiliary table.

You may also choose to standardize the width of some moldings and make specific auxiliary tables for each profile.

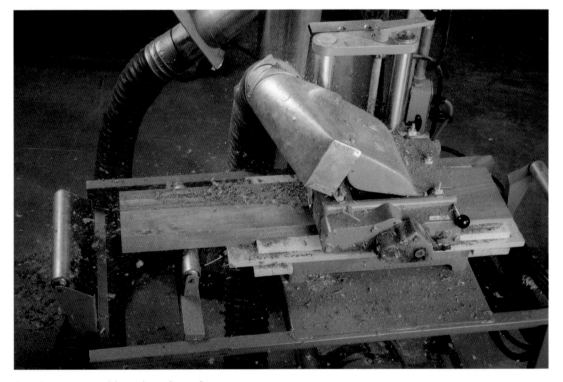

Running crown moldings through machine.

Crown moldings ready to be back cut.

Back cutting crown molding with a table saw.

Crown molding after back cutting is complete.

Using a square on both flats helps check for properly angled cuts.

Routers

You can use routers in many different ways. For example, mounted to the underside of a table, they perform the jobs of a light duty shaper. Possibly, the greatest advantage of the router is that the tool can be moved to the workpiece to perform jobs such as mortising, dadoing, and fluted pilasters. Tooling for the router comes in a wide variety of shapes and sizes. Also, the lower cost of both machine and tooling make it available to those on a limited budget.

A simple router table set up for beading shelf nosing, consisting of a vertical fence clamped to the router table, and two hold downs – one, a horizontal curved board, and the other a featherboard made from a block of wood with the end band sawed into little strips.

Shelf nosing after being beaded on a router.

Fluting a pilaster with a router and 0.375" box core bit.

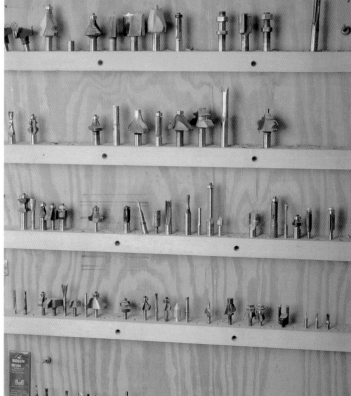

A very diverse selection of router bits exists.

Table Saws

Although not normally thought of as tools for making moldings, you can use table saws with cutter heads and auxiliary fences specifically designed for them to make moldings quickly and efficiently. Also, you can use an auxiliary fence with a normal saw blade to cut cove moldings. However, only experienced woodworkers should perform this particular operation, due to the potential danger involved.

This tool's advantage is that most shops already have one, making the cutters the only additional cost. However, table saw cutters are limited to sizes of one inch or less.

A molding head designed for a table saw. This particular one has been modified to fit the 1" arbor of our table saw.

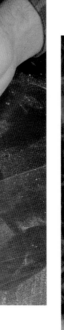

In this particular operation, we use only the first bead of the three, so we must clamp an auxiliary fence to the table saw fence to avoid cutting the good fence. Note that the clamp must be high enough to let the workpiece slide underneath.

Adjusting the auxiliary fence to cover all but one bead on a three bead cutter.

You can cut large cove moldings on a table saw by running them diagonally across the blade. Use trial and error to find the correct angle of cut. Remove stock a little at a time to avoid overworking the table saw. Use extreme caution for this molding method.

Hand Planes

Hand planes have been used throughout history to make almost every existing molding. The disadvantage, of course, is the hand plane's slowness. Further, acquiring the skill necessary to use them also requires considerable time. The advantage of hand planes is that they do an excellent job, are inexpensive, and bring users a certain satisfaction and connection to the past unachievable with any other method. Whichever equipment you decide to use, remember that designing and making your own high-quality, authentic eighteenth century moldings is one of the most important aspects of creating a great room.

You can still find hand planes for certain molding profiles, and although a bit slow, they work quite well.

When working with all power tools, make sure to either unplug or lock out power to the tool before making adjustments or touching the cutters.

Chapter Seven
Project Overview

The initial meeting with the customer is very important. You should be as prepared as possible, bringing photography of past jobs and a few books with pictures of eighteenth century rooms to show the client. Such items help to establish your client's confidence, thus accomplishing this first meeting's purpose. The client should feel completely comfortable with your ability to design and construct a beautiful room before any serious design work begins.

To begin the design process, first gain a clear understanding of how the room will be used. Today, many rooms perform multiple functions. With the advent of in-home offices, rooms are often used as business centers during the daytime – complete with computer and file storage – and TV entertainment areas in the evening hours. Your challenge will be to address all of the customer's needs for the room without losing the integrity of the design.

Picking the style for a room is usually fairly simple. Generally speaking, the room's style should reflect the overall style of the house, and the room's importance should determine the level of detail it will have. For example, a parlor or dining room usually has a much higher style of millwork than a bedroom or kitchen.

Also in one of the early meetings, the client should choose whether to use paint or a wood finish, though most customers make this decision before calling a cabinetmaker. If this decision has not been made, however, you should point out that most eighteenth century rooms were painted, so that if the customer desires a faithful reproduction room, they should choose paint also.

If the project will be painted, the wood choice is easy. Simply choose the most stable, closed-grain, cost-effective wood available in your area. In the mid-Atlantic states, we often use tulip poplar for cabinet frames and moldings because it is inexpensive and receives paint very well. For cabinet doors, cabinetmakers often use soft maple because of its stability. Using those two woods on the same job maximizes the benefits of each.

If the client chooses a wood finish, you may expedite the decision-making process simply by providing samples of each wood under discussion.

After all of these decisions have been made, the actual design and drawing processes begin. To illustrate these, we have chosen a library/office.

In an existing home, the design process may begin with actual measurements taken before any drawings are done. However, in new construction, time constraints often compel cabinetmakers to draw directly from the architect's blueprints long before the home is built. These early drawings provide important information concerning the overall construction budget. However, discrepancies between blueprints and actual measurements make this process less efficient.

Design Process: a Walnut Library

Like many rooms, this one had to perform a variety of functions. Primarily, it would serve as a combination office and library, containing a desk, a computer, a television, and shelf space adequate for hundreds of books. Secondarily, the customer would use the room to entertain guests. For this purpose, the client's only requirements were that it have a warm, yet stately atmosphere, and that it be Georgian design to match the house.

As often happens, the customer was enthusiastic about the room and brought many good design ideas. The challenge thus became integrating good eighteenth century design with the functional requirements of twentieth century living.

Once you and the client have established the overall level of detail and style, begin the design process by problem solving each wall individually. Because wall B presented the most complexity, we started there.
Wall B

Wall B had two French doors and a large fireplace which protruded into the room about 30". The depth of the fireplace created an obvious location for two deep flanking cabinets, housing computer equipment on one side and the television on the other. This design also allowed us to create great looking paneled jambs surrounding the doors.

Once we decided upon the functional layout, we began to design the area immediately surrounding the fireplace. A fireplace's level of detail and style generally sets the mood for the entire room. Often when designing a room, we try to work in some details that have personal meaning and significance for a customer. This particular homeowner was a U.S. senator, so we opted to carve a replica of the senate seal on the center block, and wheat sheaves (representing the original 13 colonies) on the pilaster blocks. This client's taste leaned strongly toward the Federal period.

Wall C and Wall A

Walls C and A were fairly straightforward in design. We carried the same details as those on Wall B around these two walls. Perhaps the only real decision we needed to make concerned the size of the raised panels. Generally speaking, try to make panels as similar to each other in width as possible. If you choose to use one-piece matched panels stock, availability will often dictate panel width.

Wall D

Wall D almost designed itself. The location and size of the door openings were fixed due to the other side of the wall's requirements; however, the symmetry of their placement made it easy to divide the bookcases into similar-sized units.

On a wall like this one, the books are the focal point: Any architectural details should frame the books, rather than call attention away from them. Therefore, we simply continued to mimic the other walls' details here, and installed a pair of handsomely matched doors.

In the balance of this chapter, we will continue through the room's installation step-by-step.

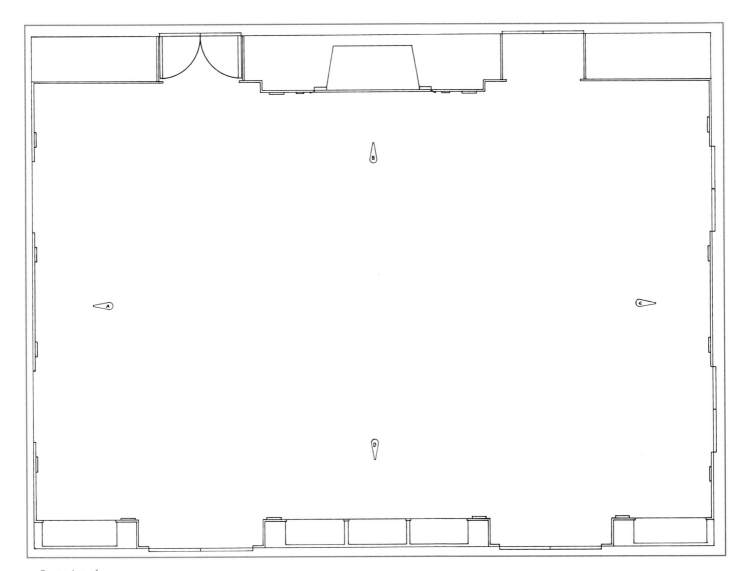

Footprint of room

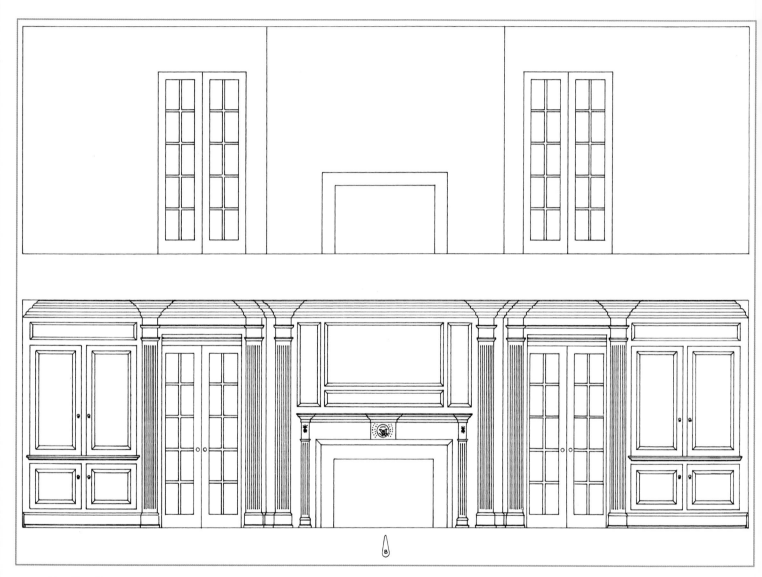

Elevation B

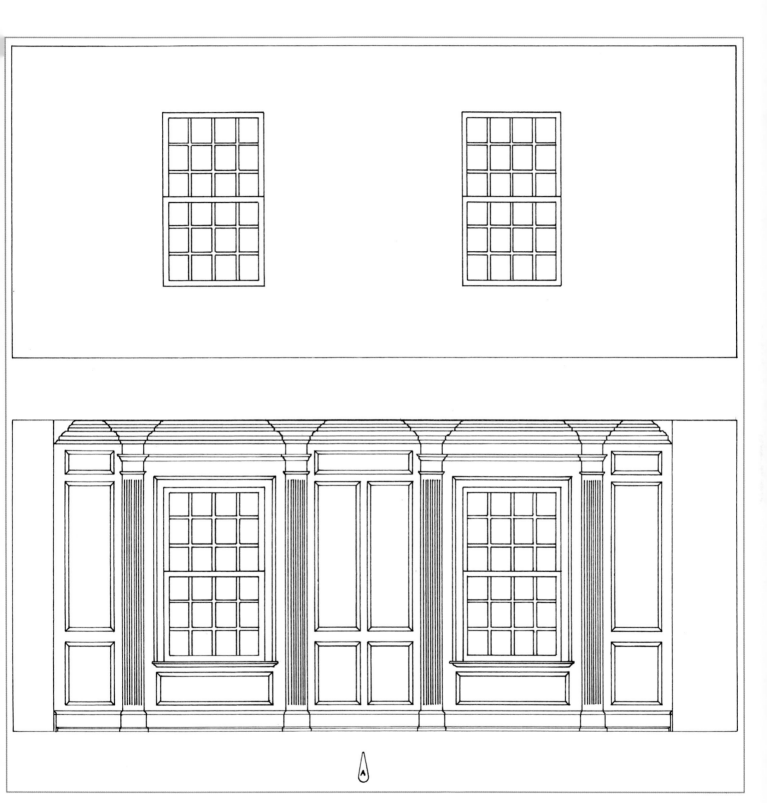

Elevation A

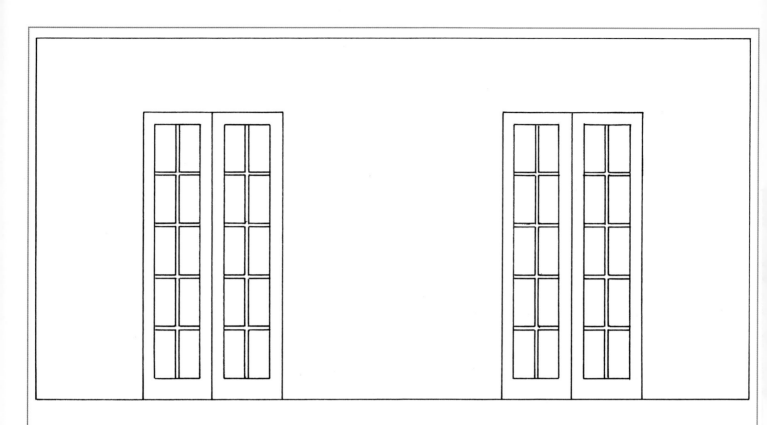

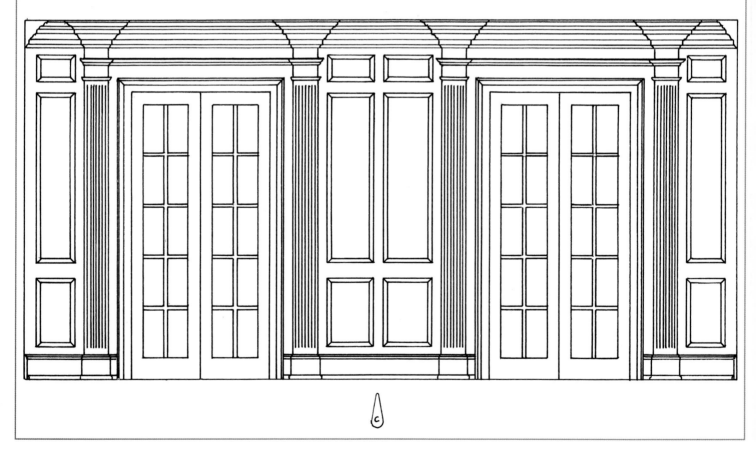

Elevation C

Elevation D

Making shop drawings based on architect's house plans.

View of fireplace wall.

Checking actual dimensions against shop drawings.

Marking story sticks.

Installation

To begin room installation, check all the walls' dimensions against your drawings. Any differences should be noted directly on the working drawings to avoid later confusion. If dimensional changes make new drawings necessary, redraw at this time.

Next, accurately mark all the features (windows, doors, fireplaces, etc.) on story sticks. Story sticks have been used by cabinetmakers, furniture makers, and boat builders for centuries and remain a very simple, inexpensive, and accurate way to transfer information from job site to shop. Lay these sticks, usually made of scrap rippings or strips of plywood, end to end around the room, and clearly mark each window jamb, door jamb, fireplace opening, heating vent, etc.

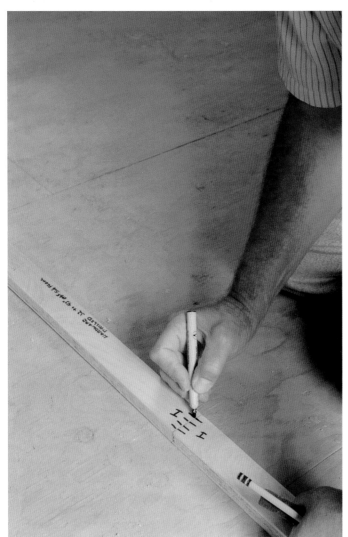

Marking story sticks.

Marking window
on story stick.

Marking door on story stick.

Selecting boards
for panel stock.

Before you build boxes and other parts of the room, give some attention to the sequence in which you will assemble the room's parts. For example, does the panel that meets the cabinet in the corner need to be installed first and the cabinet butted against it, or does the box go in first? In this example, we encountered another common problem, one concerning the way to first get the boxes into the room. Does a large box fit through the door? If not, cabinets and/or panels need to be brought in to the room in pieces and assembled on site. You must address such details before you invest effort and materials in the project. Other problems may arise, varying from job to job, such as how to bend panels around curved walls, or how to adapt panels that would cover a cold air return. Sometimes you must make structural allowances for large televisions. It's a good idea to look at your completed drawings and story sticks, make a list of problem spots, and address them.

When you have dealt with your potential problem spots, you may begin the process of making cabinets, paneling, and moldings as outlined in earlier chapters.

During the in-workshop phase of the job, make sure that you complete everything you can before installation. It's almost always more economical to work at your shop, where you have easy access to tools, than to improvise at the job location.

Some time before installation, find out whether or not the job site will be ready for you when the installation day comes. A number of other installations need to occur on location before the woodwork, including:

Plumbing: Make sure that all pipes, heating ducts, cold air returns, etc. are where agreed upon and that the pipes are long enough to reach into your cabinets if necessary.

Electric: Wires must be long enough to reach their proper places in the cabinets; find out where holes for electrical outlets need to be placed. Also, in today's high-end rooms, many electrical items have to be installed, such as security systems, fax machines, computers, televisions, stereos, thermostats, and telephone lines. Make sure that enough electrical amperage exists to run your tools.

Before attaching anything to the wall, you should make a plan of action. Figure out ahead of time all of the idiosyncrasies of the ceiling and floor. If the floor slopes down 0.75" over its length, or one wall has a large bulge in it, you must take it into account. Making a cornice mock-up (a short length of put-together cornice) helps to determine how high on the wall to place cabinets and paneling. Above all, it's important to have cabinets level. Try to avoid following the floor or ceiling unless absolutely necessary. A well-planned, level placement of cabinets initially can save you hours of work later when you trim the room.

As you begin to install pieces, make very visible marks telling you where to find the studs and joists. Assemble the pieces in your predetermined order, and fasten them to the wall. Shims serve as great helps to keep pieces level and plumb. You may cut off excess length after you have permanently attached any part in question.

Sometimes when you nail up trim or panels, no studs are available. To solve this problem, toggle bolt and glue a block of wood for nailing behind the cornice or cabinet part. (This is one advantage to having cornice moldings and baseboard, as you may easily make up discrepancies between cabinet levels and ceiling and floor levels by adjusting the trim reveals.)

When you have assembled all of the room's parts and are satisfied with the results, sand off all marks and dents in the woodwork, fill all nail holes, and prepare the room for finishing.

Marking out panels.

Marking rails and stiles.

Marking rails and stiles.

Stripping room to accept paneling.

Applying paneling around door.

Preparing panel section to accept electrical outlet before installation.

Pulling wires through paneling before installation.

133

Sliding panel section into position.

Fastening panel section.
Note: fasteners will be
hidden by pilaster.

Attaching blocking to ceiling to anchor computer cabinet.

Locating and pulling wiring through cabinet before installation.

Positioning television cabinet.

Scribing television cabinet
to exactly match paneling.

Positioning bookcases.

Installing tenons to ensure a strong joint.

Gluing tenons before assembly.

Drawing cabinets together with long bar clamps. *Note rectangular hole in top of center cabinet. In this opening, we will fasten a walnut grill for the cold air return.*

138

Cutting blocking from secondary wood to support base under pilaster strip.

Installing blocking to receive pilaster.
Note: Use secondary wood whenever
it will be completely covered by the
primary wood.

Positioning the pilaster.

Checking for proper positioning and fastening the pilaster.

Back in the shop, completing finishing touches on the fireplace surround.

Partially assembled fireplace surround.

Installed fireplace surround.
Note: Fasteners will later be
covered by carved blocks.

Bringing moldings into room for installation.

Plywood completely covering floor so that scaffolding will not do damage.

Attaching blocking
to receive cornice.

Close-up of partially
constructed cornice.

Applying third element.

Crown element being nailed to cornice.

Applying glue to biscuit to strengthen window trim joint.

Placing biscuit in slot in window trim.

Putting window into position.

Fastening window trim to window jamb with nails.

Vertical piece of backband completes the window trim.

Positioning backband head to cover joint between window trim and paneling.

Pulling wires through baseboard.

Installing outlets in baseboard.

Positioning baseboards.

Nailing baseboards.

Applying baseboard cap.

Finished fireplace wall.

Finished bookcases.

Entry doors to library.

Fireplace carving detail.

Fireplace carving detail.

Chapter Eight
Finishing Touches

When you've completed all the designing, construction, and installation, two choices that will dramatically affect the overall look of the job remain – choice of hardware and finish.

Hardware

Hardware selection for any project is simple: Use high-quality, faithful reproduction hardware that works well with the cabinetry's period and level of style. Anything else will diminish the overall look of the project. The cost of good reproduction hardware (about $5 – 7 per piece) is so minor compared to the rest of the cabinetry that it hardly makes a difference in the overall cost of the project.

Many hardware purveyors are not familiar enough with period design to be helpful to a cabinetmaker, so it is important to pick one who is knowledgeable and can give advice on period and style appropriateness as it applies to your project. (See appendix.)

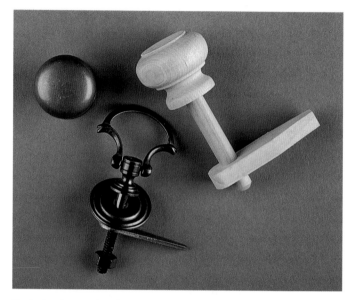

Typical cabinetry hardware. *Photo courtesy of Horton Brasses, Inc.*

Finish

The two distinct types of finishes on cabinetry are paint and clear finish such as shellac, lacquer, or varnish.
Paint

Preparation for paint finishes usually means sanding to #120 grit and filling all nail holes in trim with a non-shrinking filler. Begin the paint finish with a good sanding sealer and two coats of enamel, sanded to #220 grit between coats. A great paint job is not "rocket science," but it does require skill, experience, and conscientiousness. Be certain that you feel comfortable with your painter's ability.

Clear finish

Many woodworkers are capable of applying stain and lacquer, but to achieve a truly rich, warm finish, we usually employ the services of a professional finisher. (See appendix.)

A professional often begins coloring with a blend of aniline dyes, over which he or she applies a light-bodied pigmented oil stain. The finish then usually consists of either orange or garnet shellac, which he or she rubs out and waxes. Often, the finisher will glaze in between coats to achieve an aged look. The final result is a finish that does justice to your process of design, wood selection, and craftsmanship.

Appendix

Reference Books

Chamberlain, Samuel and Narcissa G. *New England Rooms: 1639 – 1863*. Stamford, Connecticut: Architectural Book Publishing Co., 1993.

Faulconner, Anne M. *The Virginia House*. Atglen, Pennsylvania: Schiffer Publishing, Ltd., 1984.

Garrett, Wendell. *American Colonial: Puritan Simplicity to Georgian Grace*. New York: The Monacelli Press, Inc., 1995.

Historic Houses of Philadelphia. The Barra Foundation, Inc. Philadelphia: University of Pennsylvania Press, 1998.

Howells, John Mead. *Architectural Heritage of the Piscataqua*. Washington, D.C.: publisher?, 1988.

Illustrated Dictionary of Historic Architecture. Cyril M. Harris, ed. New York: Dover Publications, date?

Raymond, Eleanor. *Early Domestic Architecture of Pennsylvania*. Atglen, Pennsylvania: Schiffer Publishing, Ltd., 1977.

Schuler, Stanley. *Architectural Details from Old New England Homes*. Atglen, Pennsylvania: Schiffer Publishing, Ltd., 1987.

Sweeney, John A. H. *The Treasure House of Early American Rooms*. New York/London: W.W. Norton and Company, 1963.

Wood Certification

Certified Forest Products Council, 14780 South West Osprey Drive, Suite 286, Beaverton, OR 97007-8424, tel. 503 590-6600, www.certifiedwood.org

Wholesale Distributors (East Coast)

TBM Hardwoods, 100 Filbert St., Hanover, PA 17331, tel. 1 800 233-5137

O'Shea Lumber, 11425 Susquehanna Trail, Glen Rock, PA 17329, tel. 1 800 638-0296, anton@oshea.com

Moisture Meters

Lignomat USA, Ltd., P.O. Box 30145, Portland, OR 97294, tel. 800 227-2105, www.lignomat.com

Sources for Specialty Lumber Sales

Good Hope Hardwoods, 1627 New London Road, Landenberg, PA 19350, tel. 610 274-8842, www.goodhope.com

Groff and Groff Lumber, 858 Scotland Road, Quarryville, PA 17566, tel. 800 342-0001

Hearne Hardwoods, 200 Whiteside Drive, Oxford, PA 19363, tel. 888 814-0007, www.hearnehardwoods.com

Irion Lumber, P.O. Box 954, Wellsboro, PA 16901, tel. 570 724-1895, www.irionlumber.com

High-quality Used Woodworking Machinery

C.S. Machinery, 210 Lower Hopewell Road, Oxford, PA 19363, tel. 610 998-0544

Tooling Sources

Inexpensive router bits and shaper cutters

Grizzley Industrial, 2406 Reach Road, Williamsport, PA 17701, tel. 800 523-4777, www.grizzley.com

Custom lockedge and molder knives

The Cayce Company, 221-B Cockeysville Road, Hunt Valley, MD 21030, tel. 800 875-0213, www.cayceco.com

Custom shaper cutters and molder knives

Charles G.G. Schmidt and Company, 301 West Grand Avenue, Montvale, NJ 07645, tel. 1 800 724-6438, www.cggschmidt.com

Reproduction Hardware

Horton Brasses, Inc., P.O. Box 95, Cromwell, CT 06416, tel. 1 800 754-9127, www.horton-brasses.com

The Coldren Company, 100 Race Street, Northeast, MD 21901, tel. 410 287-2082

Professional Finishing Services

Franklin J. Rebalsky, 59 North Main Street, Phoenixville, PA 19460, tel. 610 983-3864

Grading Rulebook

NHLA, P.O. Box 34518, Memphis, TN 38184